ENVIRONMENTAL REMEDIATION TECHNOLOGIES, REGULATIONS AND SAFETY

ASSESSING THE HYPOXIA THREAT IN US COASTAL WATERS

ENVIRONMENTAL REMEDIATION TECHNOLOGIES, REGULATIONS AND SAFETY

Additional books in this series can be found on Nova's website under the Series tab.

Additional E-books in this series can be found on Nova's website under the E-book tab.

ENVIRONMENTAL REMEDIATION TECHNOLOGIES, REGULATIONS AND SAFETY

ASSESSING THE HYPOXIA THREAT IN US COASTAL WATERS

WILLIAM JACOBS
AND
STEVE FREDERICKSON
EDITORS

Nova Science Publishers, Inc.
New York

For permission to use material from this book please contact us:
Telephone 631-231-7269; Fax 631-231-8175
Web Site: http://www.novapublishers.com

Additional color graphics may be available in the e-book version of this book.

Library of Congress Cataloging-in-Publication Data

ISBN: 978-1-62081-303-4

Published by Nova Science Publishers, Inc. †New York

CONTENTS

PREFACE

The occurrence of hypoxia, or low dissolved oxygen, is increasing in coastal waters worldwide and represents a significant threat to the health and economy of our Nation's coasts and Great Lakes. This trend is exemplified most dramatically off the coast of Louisiana and Texas, where the second largest eutrophication-related hypoxic zone in the world is associated with the nutrient pollutant load discharged by the Mississippi and Atchafalaya Rivers. Aquatic organisms require adequate dissolved oxygen to survive. The term "dead zone" is often used in reference to the absence of life from habitats that are devoid of oxygen. The inability to escape low oxygen areas makes immobile species, such as oysters and mussels, particularly vulnerable to hypoxia. This book examines the progress in understanding the dynamics of hypoxia where it occurs; understanding monitoring nutrient fluxes in watersheds; and how to reduce nutrient transport across the landscape.

Chapter 1 - Understanding and managing hypoxia requires research and management actions to address entire watersheds and their coastal receiving waters. This report examines progress in: 1) understanding the dynamics of hypoxia where it occurs (i.e., in estuaries, coastal waters, and the Great Lakes); 2) understanding and monitoring nutrient fluxes in watersheds; and 3) understanding how to reduce nutrient transport across the landscape. Because responsibility for management of nutrient enrichment in watersheds is shared across several Federal agencies, coordination and information-sharing are critical. Presently, interagency collaboration is most extensive and effective in places where investments in scientific and management activities have been largest and sustained for the longest period of time (e.g., Chesapeake Bay).

Chapter 2 - This is written testimony of Robert E. Magnien, Director, Center for Sponsored Coastal Ocean Research, US Department of Commerce.

Chapter 3 - This is testimony of Richard Greene, Chief, Ecosystem Dynamics and Effects Branch, Gulf Ecology Division, US Environmental Protection Agency.

In: Assessing the Hypoxia Threat in U.S. Coastal Waters ISBN 978-1-62081-303-4
Editors: William Jacobs and Steve Frederickson

Chapter 1

SCIENTIFIC ASSESSMENT OF HYPOXIA IN US COASTAL WATERS*

Committee on Environment and Natural Resources

COUNCIL ON ENVIRONMENTAL QUALITY OFFICE OF SCIENCE AND TECHNOLOGY POLICY EXECUTIVE OFFICE OF THE PRESIDENT

Dear Partners and Friends in our Ocean and Coastal Community,

We are pleased to transmit to you this report, *Scientific Assessment of Hypoxia in US Coastal Waters*. This document assesses the problem of hypoxia (or low dissolved oxygen) in our Nation's coastal ocean and estuarine waters. It also describes recent advances made by Federal agencies to improve scientific understanding of hypoxia and our ability to manage and, ultimately, prevent these events. In December 2004, Congress reauthorized the Harmful Algal Bloom and Hypoxia Research and Control Act (HABHRCA) by passing the Harmful Algal Bloom and Hypoxia Amendments Act of 2004. The reauthorization of HABHRCA acknowledged that hypoxia is one of the most scientifically complex and economically damaging coastal issues challenging our ability to safeguard the health of our Nation's coastal ecosystems.

This document was prepared by the Interagency Working Group on Harmful Algal Blooms, Hypoxia, and Human Health, which was chartered through the Joint Subcommittee on Ocean Science and Technology of the National Science and Technology Council and the Interagency Committee on Ocean Science and Resource Management Integration. This report complements and expands upon water quality-related priorities identified in *Charting the Course for Ocean Science in the United States for the Next Decade: An Ocean Research Priorities Plan and Implementation Strategy,* by the Joint Subcommittee on Ocean Science and Technology. It draws from the direct contributions of Federal agencies as well as

* This is an edited, reformatted and augmented version of a Committee on Environment and Natural Resources publication, dated September 2010.

previous reports and planning efforts that involved numerous experts and stakeholders from Federal, state, and local governments, academia, industry, and nongovernmental organizations. The Nation's coastal waters are vital to our quality of life, our culture, and the economy. Therefore, it is imperative that we move forward to better understand and prevent hypoxia events, which threaten all of our coasts.

This report is an effort to assess the extent of efforts to understand and lessen hypoxia events and to identify opportunities for charting a way forward. We hope it will be useful to the Congress and a broad range of interested parties.

Sincerely,

Nancy H. Sutley
Chair
Council on Environmental Quality

Sincerely,

John Holdren
Director
Office of Science and Technology Policy

JOINT SUBCOMMITTEE ON OCEAN SCIENCE AND TECHNOLOGY (JSOST)

Steve Murawski, DOC/NOAA, Co-Chair
Tim Killeen, NSF, Co-Chair
Jerry Miller, OSTP, Co-Chair

Arctic Research Commission
John Farrell
Department of Agriculture
Louie Tupas

Department of Commerce
National Oceanic and Atmospheric Administration
Craig McLean
Steve Murawski
Roger Parsons

Department of Defense
US Army Corps of Engineers Charles Chesnutt
Joan Pope

Department of Defense
Office of Naval Research Linwood Vincent
James Eckman

Department of Energy
Office of Science

Julie Carruthers
James Ahlgrimm

Department of Health and Human Services
Centers for Disease Control and Prevention
Lorraine Backer
G. David Williamson

Department of Health and Human Services
Food and Drug Administration
Robert Dickey William Jones

Department of Health and Human Services
National Institutes of Health
Allen Dearry

Department of Homeland Security
US Coast Guard
Jonathan Berkson

Department of the Interior
Kameran Onley
Tim Petty

Department of the Interior
Minerals Management Service
James Kendall
Walter Johnson

Department of the Interior
United States Geological Survey
John Haines
Linda Gunderson

Department of Justice
Matt Leopold
Amber Blaha

Department of State
David Balton
Liz Tirpak

Department of Transportation, Maritime Administration
Richard Corley

US Environmental Protection Agency
Kevin Teichman
Steven Hedtke

Executive Office of the President
Council on Environmental Quality
Hardy Pearce

Executive Office of the President
Domestic Policy Council
Paul Skoczylas

Executive Office of the President
Office of Management and Budget
Stuart Levenbach
Kimberly Miller

Executive Office of the President
Office of Science and Technology Policy
Jerry Miller

Joint Chiefs of Staff
Robert Winokur
Nadeem Ahmad

National Aeronautics and Space Administration
Jack Kaye
Eric Lindstrom

National Science Foundation
Tim Killeen
Julie Morris
Phil Taylor

Marine Mammal Commission
Robert Gisiner
Tim Ragen

Smithsonian Institution
Leonard Hirsch

JSOST Interagency Working Group on Harmful Algal Blooms, Hypoxia and Human Health (IWG-4H)

Lorraine C. Backer (Co-Chair)
Centers for Disease Control and Prevention

Paul A. Sandifer (Co-Chair)
National Oceanic and Atmospheric Administration
Paula Bontempi
Alternate: Fredric Lipschultz
National Aeronautics and Space Administration

Herbert T. Buxton
United States Geological Survey

David Garrison
National Science Foundation

Rob Magnien
Alternate: Quay Dortch
National Oceanic and Atmospheric Administration

Steven Plakas
US Food and Drug Administration

Tim Ragen
Alternate: Bob Gisiner
Marine Mammal Commission

Teri Rowles
National Oceanic and Atmospheric Administration

Joyce Saltsman
US Food and Drug Administration

Juli Trtanj
National Oceanic and Atmospheric Administration

Frederick L. Tyson
National Institute of Environmental Health Sciences

Usha Varanasi
Alternate: Walton Dickhoff
National Oceanic and Atmospheric Administration

William Russo
US Environmental Protection Agency

Mark Walbridge
Department of Agriculture

Scientific Support Staff:

Elizabeth B. Jewett
Cary B. Lopez
Carolyn Sotka
Virginia Fay
National Oceanic and Atmospheric Administration

Cheryl L. Fossani
National Science Foundation

Primary Authors

Elizabeth B. Jewett,
Cary B. Lopez,
David M. Kidwell,
Suzanne B. Bricker
National Oceanic and Atmospheric Administration

Marianne K. Burke,
Mark R. Walbridge
US Department of Agriculture

Peter M. Eldridge,
Richard M. Greene,
James D. Hagy III
US Environmental Protection Agency

Herbert T. Buxton
US Geological Survey

Robert J. Diaz
Virginia Institute of Marine Science

Major Contributors

Cheryl Brown
US Environmental Protection Agency

Bill Peterson
National Oceanic and Atmospheric Administration

Jay Peterson and Cheryl Morgan
Oregon State University

List of Acronyms

ADMS	Agricultural Drainage Management Systems
ARS	Agricultural Research Service, USDA
BMP	best management practice
CBEMP	Chesapeake Bay Environmental Model Package
CEAP	Conservation Effects Assessment Project
CENR	Committee on Environment and Natural Resources
CHRP	Coastal Hypoxia Research Program, NOAA
CRP	Conservation Reserve Program, USDA
CSREES	Cooperative State Research, Education, and Extension Service, USDA
DOD	US Department of Defense
DOE	US Department of Energy
DOI	US Department of the Interior
EISA	Energy Independence and Security Act of 2007
EMAP	Environmental Monitoring and Assessment Program
EMVL	Environmental Modeling and Visualization Laboratory
EPA	US Environmental Protection Agency
ERS	Economic Research Service, USDA
FSA	Farm Service Agency, USDA
FVCOM	Finite Volume Community Ocean Model
GLNPO	Great Lakes National Program Office, EPA
HAB	harmful algal bloom
HABHRCA	Harmful Algal Bloom and Hypoxia Research and Control Act
HCDOP	Hood Canal Dissolved Oxygen Program
IFYLE	International Field Years on Lake Erie
IOOS	Integrated Ocean Observing System
IWG-4H	Interagency Working Group on HABs, Hypoxia, and Human Health
LaMP	Lakewide Management Plan, Lake Erie
LISS	Long Island Sound Study
LMAV	Lower Mississippi Alluvial Valley
LUMCON	Louisiana Universities Marine Consortium
MARB	Mississippi Atchafalaya River Basin
MDA	Maryland Department of Agriculture
NASA	National Aeronautics and Space Administration
NAWQA	National Water-Quality Assessment Program, USGS
NBEP	Narragansett Bay Estuary Program
NCER	National Center for Environmental Research, EPA

NCII	Nutrient Control Implementation Initiative
NEP	National Estuary Program, EPA
NERL	National Exposure Research Laboratory, EPA
NERRS	National Estuarine Research Reserve System, NOAA
NHEERL	National Health and Environment Effects Research Laboratory, EPA
NMFS	National Marine Fisheries Service, NOAA
NOAA	National Oceanic and Atmospheric Administration, US Department of Commerce
NRC	National Research Council
NRCS	Natural Resources Conservation Service, USDA
NRL	Naval Research Laboratory, US Department of Defense
NRMRL	National Risk Management Research Laboratory, EPA
NSF	National Science Foundation
OST	Office of Science and Technology, EPA *OW* Office of Water, EPA
OWOW	Office of Wetlands, Oceans, and Watersheds, EPA
PDO	model Nitrogen-Phytoplankton-Detritus-Oxygen model
ReCON	Real-time Coastal Observation Network
REMM	Riparian Ecosystem Management Model
ReNuMa	Regional Nutrient Management Model
ROMS	Regional Ocean Model System
RZWQM	Root Zone Water Quality Model *SAV* submerged aquatic vegetation
SEAMAP	Southeast Area Monitoring and Assessment Program, NOAA
SPARROW	SPAtially Referenced Regressions On Watershed Attributes
STAR	Science to Achieve Results Program, EPA
STEWARDS	Sustaining the Earth's Watersheds Through Research, Data Analysis, and Synthesis
STORET	STOrage and RETrieval data system, EPA
SWAT	Soil and Water Assessment Tool
SWMP	System-Wide Monitoring Program
SWWRP	System-Wide Water Resources Program
TMDL	total maximum daily load
USACE	US Army Corps of Engineers, DOD
USDA	US Department of Agriculture
USFWS	US Fish and Wildlife Service, DOI
USGS	US Geological Survey, DOI
WRSIS	Wetland Reservoir Subirrigation System

EXECUTIVE SUMMARY

The Problem

The occurrence of hypoxia, or low dissolved oxygen, is increasing in coastal waters worldwide and represents a significant threat to the health and economy of our Nation's coasts and Great Lakes. This trend is exemplified most dramatically off the coast of Louisiana and Texas, where the second largest eutrophication-related hypoxic zone in the world is associated with the nutrient pollutant load discharged by the Mississippi and Atchafalaya Rivers.

Aquatic organisms require adequate dissolved oxygen to survive. The term "dead zone" is often used in reference to the absence of life (other than bacteria) from habitats that are devoid of oxygen. The inability to escape low oxygen areas makes immobile species, such as oysters and mussels, particularly vulnerable to hypoxia. These organisms can become stressed and may die due to hypoxia, resulting in significant impacts on marine food webs and the economy. Mobile organisms can flee the affected area when dissolved oxygen becomes too low. Nevertheless, fish kills can result from hypoxia, especially when the concentration of dissolved oxygen drops rapidly. New research is clarifying when hypoxia will cause fish kills as opposed to triggering avoidance behavior by fish. Further, new studies are better illustrating how habitat loss associated with hypoxia avoidance can impose ecological and economic costs, such as reduced growth in commercially harvested species and loss of biodiversity, habitat, and biomass. Transient or "diel-cycling" hypoxia, where conditions cycle from supersaturation of oxygen late in the afternoon to hypoxia or anoxia near dawn, most often occurs in shallow, eutrophic systems (e.g., nursery ground habitats) and may have pervasive impacts on living resources because of both its location and frequency of occurrence.

Although coastal hypoxia can be caused by natural processes, a dramatic increase in the number of US waters developing hypoxia is linked to eutrophication due to nutrient (nitrogen and phosphorus) and organic matter enrichment resulting from human activities. Sources of enrichment include point source discharges of wastewaters, nonpoint source atmospheric deposition, and nonpoint source runoff from croplands, lands used for animal agriculture, and urban and suburban areas. The incidence of hypoxia has increased ten-fold globally in the past 50 years and almost thirty-fold in the United States since 1960, with more than 300 systems recently experiencing hypoxia (Diaz & Rosenberg 2008; Table 1 and Appendix III).

Diffuse runoff from nonpoint sources, such as agriculture fields, can be difficult to control, although improved production methods that reduce tillage, optimize fertilizer application, and buffer fields from waterways can mitigate water quality impairments.

Despite the use of improved production methods in recent years, agriculture is still a leading source of nutrient pollution in many watersheds due, in part, to the high demand for nitrogen-intensive crops, principally corn. Furthermore, drainage practices, including tile drainage, have brought wetlands into crop production, short-circuited pathways (such as denitrification) that could ameliorate nutrient loading, and increased the transport of nitrogen into waterways.

Atmospheric nitrogen deposition due to fossil fuel combustion has declined in many areas due to emission controls, but it remains an important source of diffuse nutrient loading. The difficulty of reducing nutrient inputs to coastal waters results from the close association between nutrient loading and a broad array of human activities in watersheds and explains the growth in the number and size of hypoxic zones.

Unfortunately, hypoxia is not the only stressor impacting coastal ecosystems. Overfishing, harmful algal blooms (HABs), toxic contaminants, and physical alteration of coastal habitats associated with coastal development are several problems that co-occur with hypoxia and interact to decrease the ecological health of coastal waters and reduce the ecological services that they can provide.

Legislative Mandates for Action

The Harmful Algal Bloom and Hypoxia Research and Control Act (HABHRCA) mandated creation of this report, which serves as a thorough update to the first scientific assessment of hypoxia released in 2003. Several other legislative drivers also influence how Federal agencies work on coastal water quality including the Clean Water Act; the Food, Conservation, and Energy Act of 2008 ("Farm Bill"); the Energy Independence and Security Act of 2007; and the Coastal Zone Management Act. Responsibility for resolving hypoxia spans several Federal agencies (US Department of Agriculture, US Geological Survey, US Environmental Protection Agency, and National Oceanic and Atmospheric Administration), which oversee research and management/control programs (Appendix I). States play a critical role in monitoring and managing hypoxia, but their efforts are not addressed in detail here because this report was mandated to focus on Federal efforts.

Adaptive Management

Adaptive management recognizes that science should inform management decision-making, not only as the original questions are posed but as scientific understanding continues to develop.

Current scientific understanding of the factors contributing to hypoxia is derived from rigorous research that Federal agencies have been conducting and funding for at least 25 years. Federal hypoxia research has made significant progress in describing and quantifying the causes of hypoxia and reducing uncertainties required to confidently proceed with an adaptive management approach.

For example, the ecological mechanisms linking nutrient loading to hypoxia in the Chesapeake Bay were quantified sufficiently to justify an initial 40% nutrient reduction goal in the mid-1980s; further research and sophistication in simulation models have been used subsequently to support three additional rounds of adaptive management leading to more detailed nutrient reduction strategies.

From a national perspective, adaptive management approaches will have to be flexible enough to address differences among ecosystems (e.g., geography, level of watershed development, agricultural influence, nutrient loading, physical circulation of the waterbody) and future contingencies as they unfold. For example, it is likely that climate change will affect the incidence of hypoxia in coastal waters. A flexible management approach will enable response to climate impacts as they are realized. A flexible and meaningful adaptive management strategy for agricultural production—including biofuels— should be explored in order to ensure that agricultural products are produced in a manner that minimizes or prevents water quality impairments.

Reducing hypoxia via nutrient loading reductions will require a concerted effort by a diverse group of stakeholders. Implementing these actions will likely require organized programs to monitor, study, and manage water quality problems. Such programs have already been established for a few coastal waterbodies (e.g., Lake Erie, Gulf of Mexico, Chesapeake Bay, Long Island Sound).

In the case of other waterbodies where the problem has just recently been recognized and understood (e.g., Narragansett Bay, Rhode Island; Pensacola Bay, Florida), only nascent efforts, if any, are in place to address it.

Scientists use the box corer to sample sediments with negligible disturbance. Sediment cores are used to study sediment-oxygen-nutrient dynamics. Photo. L.Jewett, NOAA.

Federal Research Accomplishments and Opportunities for Advancement

Understanding and managing hypoxia requires research and management actions to address entire watersheds and their coastal receiving waters. This report examines progress in: 1) understanding the dynamics of hypoxia where it occurs (i.e., in estuaries, coastal waters, and the Great Lakes); 2) understanding and monitoring nutrient fluxes in watersheds; and 3) understanding how to reduce nutrient transport across the landscape. Because responsibility for management of nutrient enrichment in watersheds is shared across several Federal agencies, coordination and information-sharing are critical. Presently, interagency collaboration is most extensive and effective in places where investments in scientific and management activities have been largest and sustained for the longest period of time (e.g., Chesapeake Bay).

Responsibility for *monitoring dissolved oxygen* in most coastal and Great Lakes waterbodies lies with states or with Federal-state partnerships that utilize state monitoring programs to track water quality. For many coastal and Great Lakes waters where hypoxia occurs, fairly rigorous methods have been implemented for measuring dissolved oxygen and conveying this information to scientists and the public. However, the two largest hypoxic zones in the United States, located at the mouth of the Mississippi River and on the Oregon continental shelf, occur beyond the limits of state waters and, thus, rely on Federal support for monitoring. Monitoring informs coastal managers about water quality conditions in the ecosystems they oversee, and it helps support the development and verification of ecosystem simulation models used to guide management decisions.

Since the last scientific assessment of hypoxia was written in 2003, many computer *models* have been developed or updated to simulate ecological processes related to hypoxia in estuarine, coastal, and Great Lakes ecosystems and their watersheds. Sophisticated physical transport models, which examine ocean currents and mixing, are increasingly being coupled with water quality models to examine alternatives for managing hypoxia. These models are most effective when they reflect a strong scientific understanding of the processes that control hypoxia development. Relatively simple regression models can also be effective for testing hypotheses regarding factors that control hypoxia, especially as more long-term data on hypoxia become available. The relative simplicity of regression analyses has also made them useful tools for short-term ecological forecasting. Regression analyses have been used for annual hypoxia forecasts for the northern Gulf of Mexico and Chesapeake Bay.

Researchers retrieve the CTD rosette, used for collecting water and measuring dissolved oxygen and other oceanographic parameters, after a sampling profile is completed during rough seas. Photo: L.Jewett, NOAA.

Ecological box models, which simplify processes into broader geographic resolution, have improved greatly and have begun to incorporate bioeconomic components for assessing impacts of hypoxia on commercial and recreational fisheries. Advances have been particularly dramatic in watershed science. Watershed models now allow scientists to quantify the contribution of particular geographic regions to nutrients entering a coastal system. A suite of models connecting sources of nutrients in watersheds to the development of hypoxia in receiving waters, ideally with quantified uncertainty, is needed by resource managers to justify potentially costly actions to reduce nutrients and hypoxia. Use of a multiple model consensus, where models use different approaches and assumptions, provides one way to better inform management decisions. This approach is well known for its application to forecasting hurricane paths and predicting the effects of carbon dioxide fluctuations on climate.

Monitoring stream and river flow and the associated nutrients is critically important for evaluating nutrient sources in the watershed and modeling fluxes to coastal waterbodies. Stream and river monitoring was made more efficient, maximizing the use of existing resources, through the redesign of the National Stream Quality Accounting Network in 2007. However, it is important not only to maintain the current streamflow gauges and water quality monitoring stations, but also to increase the number of streams monitored. Similar to hypoxia monitoring, the monitoring of nutrients in rivers and streams is required for the development and use of predictive models and to determine the cause-effect relationships between activities that alter water nutrient export and the changes in water quality in the receiving water. Expanded stream monitoring, in combination with modeling efforts, will allow design of effective nutrient reduction strategies and will improve the ability to track progress toward water quality goals. Thus, effective and sustained management of hypoxia will require long-term support of stream and river water quality monitoring programs.

Research exploring *hypoxia impacts on fish and invertebrates* has shown negative effects on growth, reproduction, species composition, and ecological interactions among species. However, scaling up these impacts from individuals to entire fisheries populations or ecosystems is extremely complex and requires sophisticated ecosystem models that are generally beyond the state-of-the-science. Thus, research should be focused on developing these models and linking them to those that include more conventional water quality components. An additional modeling challenge will be resolving the impacts of hypoxia on living resources in systems affected by a suite of stressors as we move toward more comprehensive ecosystem management approaches.

Reducing nutrients from anthropogenic sources can lead to a reduction in coastal hypoxia and improvements in water quality throughout the watershed. Nutrient loads in some systems, such as Long Island Sound and Narragansett Bay, are being successfully reduced through targeting of wastewater treatment plants. However, nutrient loads from municipal and industrial facilities, defined as point sources, are more easily addressed than the diffuse sources of nutrients entering waterways as runoff from agricultural and urban lands.

Significant Federal investments and scientific effort have focused on developing efficient and effective ways to reduce nutrient runoff, including land conservation programs (which provide incentives for farmers to take marginal agricultural land out of production, invest in working lands to reduce erosion and control nutrients, and reestablish wetlands), alternative drainage systems, remote sensing to target fertilizer applications, winter cover crops, and newly formulated fertilizers.

Unfortunately, measures implemented on nonpoint sources to date have proven ineffective at reducing nutrient loads from large watersheds to levels needed to significantly reduce hypoxic zones. For example, reductions in total spring nitrogen loads to the Gulf of Mexico from 2001- 2005 were primarily from forms of nitrogen other than nitrate, which is a critical form fueling spring primary production leading to hypoxia. This highlights the importance of the nutrient composition, as well as seasonality, for reduction targets. Further, evidence suggests that some ecosystems may not respond as expected to nutrient management as they become nutrient saturated. Thus, major new initiatives should include rigorous scientific evaluation of the effectiveness of existing nutrient reduction strategies through local analyses of downstream water quality so that strategies can be refined accordingly. Continued research is also needed on new practices, such as those that enhance natural nutrient processes through reforestation, river diversions into wetlands, and vegetation buffer systems for streams and rivers. Such research has already led to management decisions to protect wetlands and to reforest portions of watersheds.

Establishment of criteria for acceptable nutrient levels in both coastal areas and in rivers upstream from hypoxic zones is also a critical tool for designing management plans. These criteria can also be used to measure progress.

Conclusion

Hypoxia is a major contributor to the decline of coastal water quality observed in recent decades, and its extent has been expanding. It is part of the broader issue of nutrient-driven eutrophication. Eutrophication is also linked to increased HABs, loss of seagrasses, and other impacts on coastal ecosystems. For eutrophic ecosystems, concerted and coupled research and management efforts, along with stakeholder support, will be needed to rigorously identify, quantify, and implement nutrient reduction strategies that are effective and achievable. Furthermore, for systems such as the Oregon shelf where hypoxia is driven primarily by natural processes linked to variations in climate, improved scientific understanding will provide insight into future impacts of climate change on similar ecosystems. Moreover, knowledge gained will be important for developing forecasts of the extent and severity of low dissolved oxygen, which will help managers mitigate the impacts of hypoxia.

If properly planned and executed, adaptive management of nutrient inputs will lead to significant reductions in hypoxia. However, if current practices are continued, the expansion of hypoxia in coastal waters will continue and increase in severity, leading to further impacts on marine habitats, living resources, economies, and coastal communities.

Sunset aboard EPA's Ocean Survey Vessel Peter W. Anderson during a research cruise in the Gulf of Mexico hypoxic zone. Photo: EPA.

1. Legislative Background, Report Overview, and Development Process

1.1. Legislative Background

In the early 1980s, concern about low dissolved oxygen in coastal waterbodies of the United States led to the first national assessment of coastal hypoxia (Whitledge 1985), which found dissolved oxygen levels in many US waterbodies on the decline as a result of eutrophication. By the 1990s, serious and large-scale water quality problems were identified, including harmful algal blooms (HABs) and hypoxia, most prominently in the northern Gulf of Mexico, Lake Erie, Chesapeake Bay, and Long Island Sound. These problems led to a national assessment of eutrophication in 1999 (Bricker et al. 1999), which was subsequently updated (Bricker et al. 2007), and an integrated assessment of hypoxia in the northern Gulf of Mexico (CENR 2000, US EPA 2007), as well as the passage of the Harmful Algal Bloom and Hypoxia Research and Control Act of 1998 (HABHRCA, Public Law 105-383).

HABHRCA was reauthorized by the Harmful Algal Bloom and Hypoxia Amendments Act of 2004 (HABHRCA 2004, Public Law 108-456). HABHRCA 2004 reconstituted the Interagency Task Force on HABs and Hypoxia. To fulfill the requirements of both HABHRCA 2004 and the Oceans and Human Health Act of 2004, the Interagency Task Force on HABs and Hypoxia was incorporated into the Interagency Working Group on HABs, Hypoxia, and Human Health (IWG4H, see page iii) of the Joint Subcommittee on Ocean Science and Technology. HABHRCA 2004 required five reports to assess and recommend research programs on HABs and hypoxia in US waters, including this report, a *Scientific Assessment of Hypoxia in US Coastal Waters* (Box 1). HABHRCA 2004 stipulates that this report should:

- examine the causes, ecological consequences, and economic costs of hypoxia;
- describe the potential ecological and economic costs and benefits of possible actions for preventing, controlling, and mitigating hypoxia;
- evaluate progress made by and needs of Federal research programs; and
- identify ways to improve coordination among Federal agencies with respect to research on hypoxia.

HABHRCA 2004 also authorizes appropriations to the Secretary of Commerce and the National Oceanic and Atmospheric Administration (NOAA) to conduct research on hypoxia.

Additional legislation affecting Federal research on hypoxia or factors that may affect hypoxia is listed in Box 2. The Coastal Zone Act Reauthorization Amendments of 1990 reauthorized the Coastal Zone Management Act and established a joint program between the US Environmental Protection Agency (EPA) and the states to develop and implement coastal nonpoint pollution management programs. We note that EPA's Mississippi River Basin program will be enhanced in 2011, pending enactment of the President's proposed budget. The Water Resources Development Act authorizes flood control, navigation, and environmental projects and studies by the US Army Corps of Engineers (USACE), and it states that the Secretary of the Army may participate in assessments of hypoxia in the northern Gulf of Mexico.

Box 1. HABHRCA 2004 Reports and Assessments

- Harmful Algal Bloom Management and Response: Assessment and Plan
- National Assessment of Efforts to Predict and Respond to Harmful Algal Blooms in US Waters (Prediction and Response Report)
- Report on National Scientific Research, Development, Demonstration, and Technology Transfer Plan for reducing HAB Impacts (RDDTT Plan)
- Scientific Assessment of Freshwater Harmful Algal Blooms
- Scientific Assessment of Marine Harmful Algal Blooms
- *Scientific Assessment of Hypoxia in US Coastal Waters*

Box 2. Legislation Relevant to Hypoxia

- Harmful Algal Bloom and Hypoxia Research and Control Act
- Coastal Zone Management Act
- Water Resources Development Act
- Clean Water Act
- Food, Conservation, and Energy Act of 2008
- Energy Independence and Security Act of 2007
- Annual appropriations of Federal agencies

Some of this legislation focuses on water quality in upland watersheds, which ultimately may affect the quality of water delivered to coastal ecosystems; therefore, it also impacts coastal hypoxia. The Clean Water Act aims to restore and maintain the chemical, physical, and biological integrity of the Nation's waters by reducing point and nonpoint pollution sources, providing assistance to publicly owned treatment works for improving wastewater treatment, and maintaining the integrity of wetlands. The Food, Conservation, and Energy Act of 2008 (the "Farm Bill") authorized $25 billion to support conservation programs administered by the US Department of Agriculture (USDA) through its Natural Resources Conservation Service (NRCS) and Farm Service Agency (FSA). Finally, the Energy Independence and Security Act (EISA) of 2007 potentially has far-reaching environmental implications. The bill mandates production of 36 billion gallons of biofuels by the year 2022, including 15 billion gallons of corn-based ethanol.

The legislation designates the EPA, USDA, and US Department of Energy (DOE) as lead agencies in an interagency effort to develop criteria for sustainable development of biofuels, which include prevention of water quality and ecosystem impairments.

1.2. Report Overview

This assessment focuses on the science that forms the basis for improving knowledge of the causes, impacts, and solutions for hypoxia in coastal ecosystems of the United States as of 2009. This report has four sections and three appendices:

- section 1: Legislative background, report overview, and report development process;

- section 2:The current status of hypoxia in US coastal waters, the spectrum of causal factors contributing to hypoxia, and the ecological and economic consequences of hypoxia;
- section 3: Status and accomplishments of current Federal monitoring, assessment, and research activities related to hypoxia, including research on coastal ecosystems and their watersheds that provides the basis for effective management actions;
- section 4: Directions for future science activities and opportunities to increase effectiveness and cost-efficiency through coordination of programs across Federal agencies;
- Appendix I: Federal Government Hypoxia or Hypoxia-related Research;
- Appendix II: Case studies that feature coastal ecosystems across the United States affected by hypoxia. The featured systems were selected to illustrate the range of circumstances causing hypoxia as well as the range in status of scientific understanding and management approaches being implemented (the case studies are referenced throughout the report); and
- Appendix III: Details on US coastal systems affected by hypoxia.

1.3. The Report Development Process

This report was prepared by a task force associated with the IWG-4H that included representatives of Federal agencies participating in the science and management of coastal hypoxia.

It builds on earlier reports to assess hypoxia in US coastal waters by updating the assessments and summarizing the major advances in hypoxia research during the past five years. Specifically, this report draws on *An Assessment of Coastal Hypoxia and Eutrophication in US Waters* (CENR 2003), which was called for in HABHRCA 1998. This report also recommends priorities for future hypoxia-related research across the US government.

2. Causes and Status of Hypoxia in US Coastal Waters

2.1. The Issue of Hypoxia

Hypoxia, or water with dissolved oxygen that is too low to support fish and other important species (Box 3), has been recognized as one of the most important water quality problems worldwide (Diaz and Rosenberg 2008). Hypoxia has become a serious problem along all of the Nation's coasts and in the Great Lakes. Hypoxic areas are often termed "dead zones" since the only organisms that thrive there are those that can live with little or no oxygen, often only bacteria. Most mobile organisms are able to avoid hypoxic waters by swimming or crawling away; organisms unable to move or those trapped within a zone of hypoxia become physiologically stressed and die if exposure is prolonged or severe (Diaz and Rosenberg 1995, Vaquer-Sunyer and Duarte 2008).

Most coastal hypoxia is associated at the global scale with either areas of high population density or developed watersheds that export large quantities of nutrients and organic matter (Figure 1, Rabalais et al. 2007, Galloway et al. 2008, Diaz and Rosenberg 2008). The number of waterbodies with recorded and published accounts of low dissolved oxygen from around the globe has increased about an order of magnitude during the last 50 years, from less than 50 in the 1960s to about 400 by 2008 (Diaz and Rosenberg 2008). The number of waterbodies in the United States with documented hypoxia followed the same trend, increasing from 12 prior to 1960 to over 300 by 2008 (Appendix III; Figure 2).

The second largest eutrophication-related hypoxic area in the world (after the Baltic Sea, which is approximately 80,000 square kilometers, or km^2, Karlson et al. 2002, Hansen et al. 2007) occurs in the United States, and is associated with the discharge from the Mississippi/Atchafalaya Rivers in the northern Gulf of Mexico. The Gulf of Mexico hypoxic area is similar in extent and volume to another large hypoxic area associated with China's Changjiang River outflow (13,700 km^2 in 1999, Li et al. 2002). The northern Gulf of Mexico hypoxic area has increased substantially in size since the mid-1980s when it was first measured at about 4,000 km^2 (Rabalais et al. 2007). In 2008, it encompassed 20,719 km^2, the second largest area on record (http://www.gulfhypoxia. net/). A concerted Federal interagency and multi- state effort has been underway since 2000 to find ways to reduce the size of the hypoxic area (CENR 2000, US EPA 2007; see Gulf of Mexico Case Study, Appendix II).

Much of the scientific interest and management concern about hypoxia is focused on its principal cause, *eutrophication*, which is defined as "an increase in the rate of supply of organic matter to an ecosystem" (Nixon 1995). Eutrophication is most often associated with nutrient enrichment of coastal waters from urban and agricultural land runoff, wastewater treatment plant discharges, and air deposition of nutrients (Bricker et al. 2007, Galloway et al. 2008) (see Section 2.3.2). The combination of stressors associated with eutrophication, which includes both HABs and hypoxia, continues to degrade US coastal waters, estuaries, and the Great Lakes. Nutrient management, primarily via improved treatment of industrial and municipal point source discharges, has reduced hypoxia in some systems (e.g., Los Angeles Harbor, Delaware River, and Hillsborough Bay in Florida). However, a concerted and sustained effort to address nonpoint sources of nutrients will be needed to allow more waterbodies to recover and to protect our estuarine and coastal resources for the future.

2.2. Status of Hypoxia in the United States

Hypoxia is a serious problem in estuarine and coastal areas of the United States resulting from a diverse set of causes, including eutrophication driven mainly by nonpoint sources of nutrients (Diaz and Rosenberg 2008). The present day status of hypoxia contrasts with that of fifty years ago when it was primarily a problem in rivers resulting from inadequate treatment of municipal and industrial wastes.

Through the 1970s, organic matter loading from sewage and industrial discharges was the most salient cause of eutrophication and hypoxia in the United States (Smith et al. 1987). The majority of the 69 hypoxic US waterbodies known from that time period were related to discharges from wastewater treatment (e.g., Providence River, Melrose et al. 2007) or industry (e.g., Pensacola Bay, Hagy and Murrell 2007).

Box 3. Hypoxia Definition

Hypoxia means “low oxygen”. In aquatic and marine systems, low oxygen generally refers to a dissolved oxygen concentration less than 2 to 3 milligrams of oxygen per liter of water (mg/L), but sensitive organisms can be affected at higher thresholds (4.5 mg/L). A complete lack of oxygen is called anoxia.

Hypoxic waters generally do not have enough oxygen to support fish and other aquatic animals, and are sometimes called dead zones because the only organisms that can live there are microbes. The criteria set for health of various species in Chesapeake Bay are a good example of how one definition for hypoxia is not possible (EPA 2003).

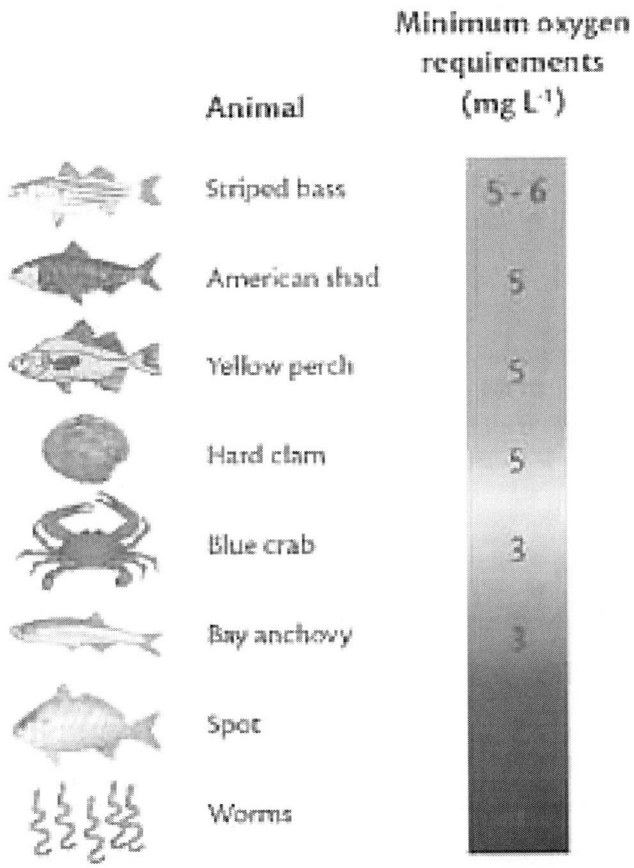

Minimum dissolved oxygen requirements for key organisms in Chesapeake Bay

Actions following the passage of the Clean Water Act in 1972 reduced these point source organic loads and, much later, began to reduce point source phosphorus and (even later) nitrogen loading, contributing to some improvements in low dissolved oxygen problems.

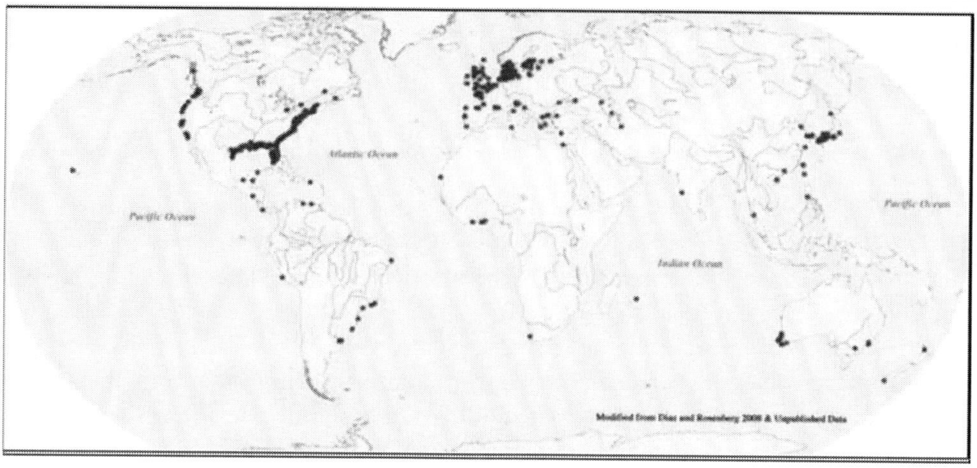

Figure 1. Global distribution of systems affected by low dissolved oxygen.

The Clean Water Act, however, did not explicitly address management and regulation of nonpoint sources of pollutants. It was in the 1970s that the rising effects of nonpoint runoff of nutrients were observed in water quality deterioration around the United States (Smith et al. 1987). The 1970s were also when first reports of hypoxia appeared for our larger coastal systems (northern Gulf of Mexico, New York Bight, Long Island Sound). By the 1980s, it became apparent that increasing nutrient loads from nonpoint sources were responsible for an expansion of eutrophic conditions in estuarine and coastal systems within the United States (reviewed by Bricker et al. 1999, 2007). Overall, the United States has led a global trend in which agricultural fertilizers are the leading nonpoint source of nutrient pollution, but atmospheric deposition from burning of fossil fuels and nitrogen fixation associated with legume crops also contribute significantly (Howarth et al. 1996). The primary sources of nutrient pollution vary substantially among watersheds (e.g., Figure 3), complicating the task of nutrient management.

A current analysis of 647 US coastal and estuarine ecosystems indicates that the percentage of systems with hypoxia has increased dramatically since the 1960s (Figure 2) and even since the 1980s (Table 1). The first national assessment of oxygen conditions in US waters, conducted in the 1980s, found 38% of systems to have hypoxia (Whitledge 1985). Subsequent eutrophication assessments found greater than 50% of systems to have hypoxia (Bricker et al. 1996, 1997a, 1997b, 1998a, 1998b, 2007). Updating the information from all these sources finds that 307 of 647 ecosystems have hypoxia (Appendix III). Most of the increases since the 1980s occurred in the North Atlantic, South Atlantic, and Pacific regions and reflect an increase in the incidence of hypoxia.

In the Middle Atlantic and Gulf of Mexico regions, the percentage of hypoxic systems was already high in the 1980s and remains high to date. Further, the third *National Coastal Condition Report* (US EPA 2008) found water quality for Atlantic and Gulf coasts to range from fair to poor, based on data from the early 2000s. For the Great Lakes, there appears to have been little change in hypoxic conditions since the 1980s.

Dissolved oxygen conditions have improved in some waterbodies due to intensive regulation of nutrient or carbon inputs, primarily from more advanced sewage and industrial treatment.

Table 1. Percentage of US estuaries and coastal waterbodies with reports of hypoxia

Region	% Hypoxic 1980s[1]	% Hypoxic 1990s[2]	% Hypoxic 2000s[3]	% Hypoxic 2000s[4]
North Atlantic	6 (1 of 17)	22 (4 of 18)	35 (7 of 20)	26 (10 of 38)
Mid-Atlantic	50 (19 of 38)	59 (13 of 22)	64 (14 of 22)	42 (76 of 180)
South Atlantic	18 (9 of 51)	64 (14 of 22)	91 (20 of 22)	55 (77 of 139)
Gulf of Mexico	69 (38 of 55)	84 (32 of 38)	68 (26 of 38)	51 (105 of 205)
Pacific	21 (6 of 29)	26 (10 of 39)	35 (13 of 37)	46 (37 of 80)
Great Lakes	1 of 5 lakes	2 of 5 lakes		2 of 5 lakes
Total Nation	38 (73 of 190)	52 (73 of 139)	58 (80 of 139)	47 (307 of 647)

[1] Based on Whitledge 1985a;
[2] Based on Bricker et al. 1996, 1997a, 1997b, 1998a, 1998b;
[3] Based on Bricker et al. 2007;
[4] From Appendix III.

The best examples are from smaller systems where point source discharges were the primary source of organic matter and nutrients, such as the Hudson River in New York and the Delaware River in Pennsylvania and New Jersey (Brosnan and O'Shea, 1996; Patrick, 1988). In larger systems where point sources have been intensely managed, dissolved oxygen conditions have not improved because of large nonpoint sources of nutrients. Examples include Chesapeake Bay and Lake Erie, which have a long history of nutrient management but continuing problems with large-scale hypoxia. (See Appendix II). Even with improved management of nonpoint sources, recovery of some large systems may be delayed due to nutrient and organic matter that have accumulated in the sediments over the years, thereby increasing oxygen demand, and, in turn, expanding the amount of hypoxia for a given load of nutrients (Turner et al. 2008). This may be the case in the northern Gulf of Mexico and Chesapeake Bay (Hagy et al. 2004, Turner et al. 2008). Unfortunately, nutrient loading from agricultural and atmospheric sources is predicted to increase globally over the next 50 years (Tilman et al. 2001, Galloway et al. 2008), creating additional stress on coastal waterbodies already in decline. Federal research and management actions must accelerate the pace of water quality management in order to protect and restore valued coastal aquatic resources.

2.3. Causes of Hypoxia

Widespread and persistent hypoxia is generally not a natural condition in estuaries, coastal waters, or large lakes like the Great Lakes, with Lake Erie being an exception (Appendix II). However, naturally occurring oxygen depletion does occur, as in oceanic oxygen minimum zones, fjords, some stratified seas (e.g., the Black Sea), lakes and ponds, and swamps and backwaters that circulate poorly and have large loads of natural land-derived organic matter. Naturally occurring hypoxia is not the focus of this report, but it is important to consider these cases because human activities may increase the frequency, duration, and intensity of naturally occurring hypoxia (Cooper and Brush 1991, Helly and Levin 2004, Diaz and Rosenberg 2008).

When the rate of oxygen consumption in aquatic environments increases such that the oxygen consumption rate exceeds resupply, oxygen concentrations can quickly decline to levels insufficient to sustain most animal life. Two conditions are generally required for the development and maintenance of hypoxia (Figure 4):

- water column stratification that isolates a layer of bottom water and sediments from a usually well-oxygenated surface layer (see Section 2.3.1), and
- a source of organic matter, which is then decomposed and depletes oxygen in the isolated bottom layer (see Section 2.3.2).

Diel-cycling hypoxia, where conditions can cycle from supersaturation of oxygen late in the afternoon to hypoxia or anoxia near dawn, is a unique situation that does not require the first condition described above.

This phenomenon can occur in *unstratified* waters and appears to be increasing in frequency (D'Avanzo and Kremer 1994, Verity et al. 2006, Shen et al. 2008). It most often occurs in shallow, eutrophic tidal creeks where oxygen exchange at the water surface is insufficient to supply the respiratory needs of microbes and other organisms.

This type of transient or diel hypoxia may cause pervasive impacts on living resources both because of the frequency with which it occurs (hours to days) and its location (small and shallow nursery ground habitats, e.g., Pepper Creek, Delaware; Tyler and Targett 2007, Tyler et al. 2008). The increased production of organic matter within a system, mostly in the form of phytoplankton productivity in response to anthropogenic nutrient pollution, contributes to both seasonal and diel-cycling hypoxia in coastal ecosystems. In shallow systems that are highly eutrophic, calm weather conditions and extended periods of cloud cover can cause low dissolved oxygen events (Tyler et al. 2008).

Whereas physical factors create the physical conditions that are usually necessary for hypoxia to develop in bottom waters, nutrient-driven eutrophication provides the increased deposition of decomposing organic matter to the bottom layer, creating increased oxygen demand.

Although physical factors can change over the long-term, they have not in most cases (e.g., Hagy et al. 2004, Greene et al. 2009). Thus, eutrophication is the principal cause of the long-term increases in hypoxia that have been observed.

2.3.1. Physical Factors

Density stratification of the water column, in which a less dense layer of water floats on top of a denser bottom layer, is an almost universal characteristic of coastal systems subject to seasonal bottom water hypoxia. Stratification reduces the potential for oxygen from the atmosphere to replenish oxygen depleted at depth. In most cases involving marine systems, a vertical gradient of salinity, creating a halocline, is the most important factor contributing to density stratification. Surface heating, creating warmer surface water temperatures and thus a vertical gradient of temperature, or a thermocline, is the cause of stratification in lakes and can also contribute to density stratification in marine systems, especially during spring when deeper waters are relatively cold. Thermal stratification creates the potential for hypoxia in Lake Erie and is also known to be important in Long Island Sound and New York Bight (Boesch and Rabalais 1991, Hawley et al. 2005, Lee and Lwiza 2008).

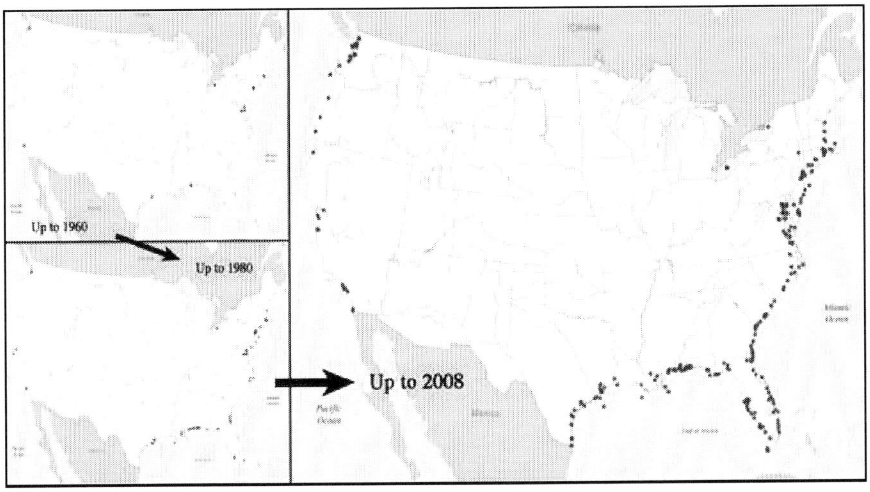

Figure 2. Change in number of US coastal areas experiencing hypoxia from 12 documented areas in 1960 to over 300 now (Appendix III). Not shown here are one hypoxic system in Alaska and one in Hawaii.

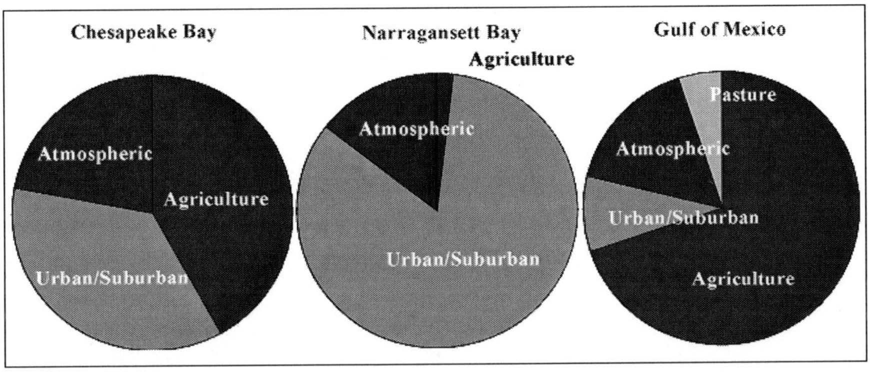

Figure 3. Comparison of the relative contribution of major sources of nitrogen pollution in three coastal ecosystems experiencing hypoxia. Urban/suburban includes both point (industrial and sewage effluent) and nonpoint sources (residential run off). Data sources: Chesapeake Bay: Chesapeake Bay Program; Narragansett Bay: Nixon et al. 2008, Moore et al. 2004; Gulf of Mexico: Alexander et al. 2008.

In the fall, cooling temperatures and decreasing freshwater inputs can destabilize summer stratification, leading to relatively abrupt mixing or "turnover" of the water column that eliminates seasonal hypoxia. Fall turnover is often associated with a wind event such as a storm or frontal passage. In river-dominated estuaries and ocean margins, density stratification and other circulation features are strongly affected by the amount of less dense freshwater flowing out of associated rivers. Therefore, the extent and severity of seasonal hypoxia varies on an annual basis in relation to the magnitude of freshwater inputs.

In larger systems, where freshwater is retained in the system for longer periods of time, summer hypoxia is usually coupled most closely to freshwater inputs during spring, when spring rains or snow melt cause high freshwater flow (e.g., Chesapeake Bay, Hagy et al. 2004; northern Gulf of Mexico, Greene et al. 2009).

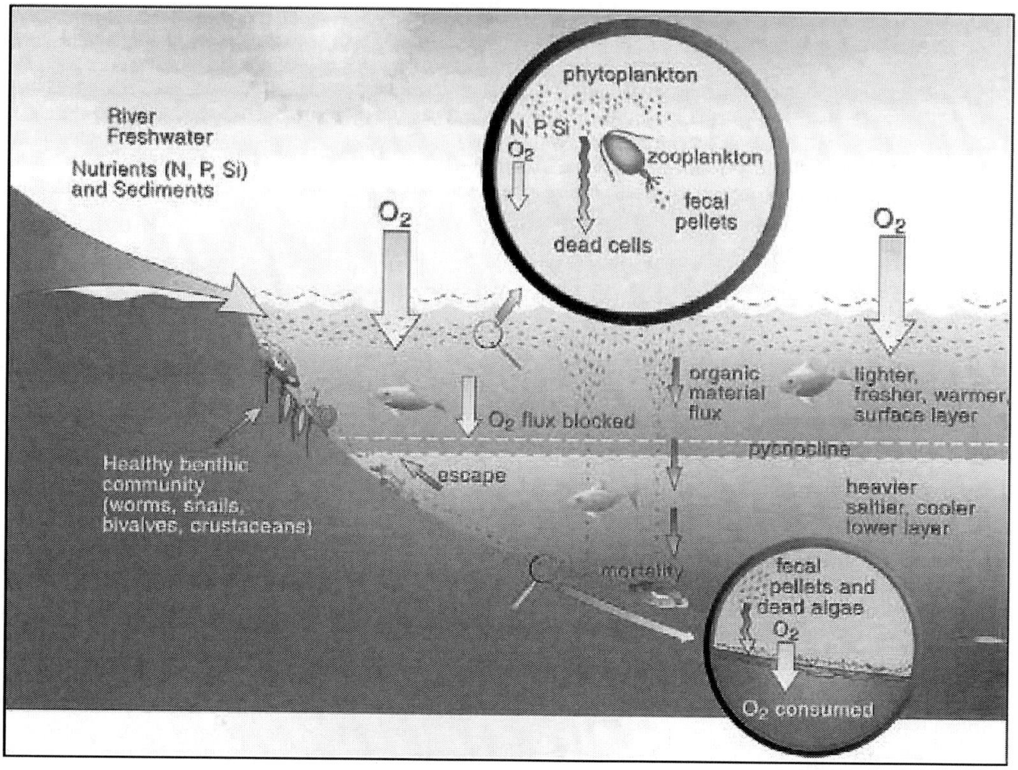

Figure 4. Conceptual diagram illustrating development and effects of hypoxia in stratified waters (from Downing et al. 1999). The pycnocline is the boundary that stratifies and separates the bottom and surface water layers.

In smaller systems, physical conditions conducive to development of hypoxia may be coupled to freshwater inputs on a shorter time scale. For example, a single early fall pulse of freshwater associated with Hurricane Ivan stimulated development of an unseasonable period of hypoxia in Pensacola Bay, Florida, during fall 2004 (Hagy et al. 2006). Shen et al. (2008) observed periods of hypoxia lasting two to five days in a shallow mid-Atlantic creek following a rain event.

Development and maintenance of hypoxia are strongly affected by water column mixing. Strong tidal mixing prevents water column stratification almost entirely in some systems. Where tides are not as strong, low dissolved oxygen events may occur during neap tides, the phase of the spring-neap tidal cycle when tidal mixing is minimal (Haas 1977). Stratification, and thus vulnerability to hypoxia, is reduced during the larger spring tides, which are associated with full and new moons, when tidal currents are stronger and generate increased mixing of the water column. Some estuaries, such as Mobile Bay, Alabama, and Pensacola Bay, Florida, in the northeastern Gulf of Mexico have low amplitude tides (Hagy and Murrell 2007), and others, such as the Albermarle-Pamlico Estuarine System on the North Carolina coast, are virtually tideless (Luettich et al. 2002). These estuaries, which are also in a warmer climate that yields higher respiration and nutrient regeneration rates, are particularly susceptible to stratification and hypoxia. Wind is an especially important factor affecting hypoxia in these microtidal estuaries (Reynolds-Fleming and Luettich 2004, Park et al. 2007, Hagy and Murrell 2007).

2.3.2. Biological/Chemical Factors

Eutrophication, fundamentally a biological and biogeochemical process, is an important factor and indeed the principal cause of coastal hypoxia. In an update to the *National Estuarine Eutrophication Assessment* (Bricker et al. 2007, See Table I), eutrophication was recognized as a widespread problem, with 65% of assessed systems (99 of 141 systems had adequate data for analysis) showing moderate- to high-level problems. The most commonly reported eutrophication-related problems included hypoxia, losses of submerged grasses, excessive algal blooms (the most commonly reported problem), and numerous occurrences of nuisance and toxic HABs. The mid-Atlantic was identified as the region most impacted by eutrophication. The majority (almost 60%) of estuaries assessed, with the exception of North Atlantic systems (Cape Cod north to Maine), were highly influenced by human-related activities that contributed to land-based sources of nutrient loads. The most commonly reported causes of nutrient-related impairments were agricultural activities (row crops and livestock operations), wastewater treatment plants, urban runoff, and atmospheric deposition. Eutrophication-related problems were predicted to worsen in 65% of estuaries, whereas 19% of the assessed estuaries were expected to improve in the future. Analysis of the extent of change in 58 estuaries from the early 1990s to the early 2000s shows that most (55%) were unchanged, including all of the larger estuaries.

The principal sources of nutrients that sustain coastal eutrophication vary among systems. Where significant rivers enter the coastal zone, nutrient loads associated with the freshwater discharge are usually the predominant source. Coastal systems lacking substantial riverine inputs may receive a substantial fraction of nutrient inputs via submarine groundwater discharge or direct atmospheric deposition, as in Barnegat Bay, New Jersey (Kennish et al. 2007). The major anthropogenic sources of nutrients to coastal waters are row crop agriculture and animal operations, industrial and municipal wastewater discharges, nonpoint source runoff from urban and suburban areas, and atmospheric deposition (Figure 3). Human activities have been linked to a dramatic increase in nitrogen cycling in the natural environment. Watersheds can become “nitrogen-saturated”, for example, when further additions of nitrogen do not elicit a response from the ecosystem. In nitrogen-saturated watersheds, the surplus nitrogen will theoretically leak to streams, groundwater, or the atmosphere (Vitousek et al. 1997). Loss or degradation of riverine and coastal wetlands can also contribute to eutrophication of coastal waters because wetlands naturally trap and retain nutrients (e.g., Valiela and Cole 2002).

In most cases, eutrophication has been caused by an anthropogenic increase in nutrient inputs. In a few cases, large but, nonetheless, natural sources of nutrients have been implicated in sustaining high primary productivity in coastal systems. For example, some watersheds in southwest Florida have naturally high concentrations of phosphates that can support high levels of primary production (Turner et al. 2006a). Natural upwelling of nutrient-rich deep ocean water into shallow areas also can support large blooms of phytoplankton (Glenn et al. 1996, Chan et al. 2008) and can result in hypoxia. Upwelling of nutrient-rich waters has been implicated in the development of severe widespread hypoxia and, for the first time (in 2006), anoxia on Oregon’s inner continental shelf (Grantham et al. 2004, Chan et al. 2008). Transport of hypoxic water from the continental shelf into coastal embayments has also been documented along the Oregon coast (Brown et al. 2007). These developments seem to be linked to impacts of climate variation on ocean processes, such as intensity of upwelling winds, oxygen concentration in upwelled water, and water column respiration (Grantham et al. 2004, Chan et al. 2008). More subtle upwelling has been observed along the New Jersey coast and has been implicated in development of nearshore hypoxia (Glenn et al. 1996, 2004). Both of these cases, however, contrast with the situation for most coastal areas undergoing eutrophication, where the source of increased nutrient input is land-based and attributable to anthropogenic causes.

Hypoxia can be related in some cases to documented long-term increases in algal biomass. For example, the long-term increase in chlorophyll-*a* (a proxy for phytoplankton biomass) in the Chesapeake Bay, as documented by Harding and Perry (1997), was concurrent with the increase in hypoxia documented by Hagy et al. (2004). In a few cases, hypoxic events have been related to high biomass HABs. For example, the one-time hypoxic event in the New York Bight (Garlo et al. 1979, Boesch and Rabalais 1991) and hypoxia in the Peace River-Charlotte Harbor system (Turner et al. 2006a) were both related to high biomass algal blooms. Further, massive mortalities of bottom-dwelling organisms were associated with the cascading effects of algal toxicity, hypoxia, anoxia, and hydrogen sulfide poisoning during a toxic HAB event in Florida in 2005 (Landsberg et al. 2009).

2.4. Consequences of Hypoxia

2.4.1. Ecological Consequences

Ecological effects of hypoxia in coastal systems can vary in both degree and scale (Figure 5). The specific concentration of dissolved oxygen below which various animals suffocate varies, but for estuarine and marine species, effects generally appear when oxygen drops below about 3 mg/L (Diaz and Rosenberg 1995, Ritter and Montagna 1999, Rabalais et al. 2001, Breitburg et al. 2001, Karlson et al. 2002). However, the behavior of some organisms (e.g., fish, larvae) can be negatively affected at oxygen concentrations as high as 4-4.5 mg/L (Whitmore et al. 1960, Kramer 1987, Breitburg 1994). Toxic hydrogen sulfide is often present in waters nearly or completely devoid of oxygen and significantly reduces survival (Gamenick et al. 1996).

Fish kills are an obvious and very unpleasant consequence of hypoxia. The frequency and magnitude of fish kills have increased as nutrient-related eutrophication has worsened both hypoxia and HABs (Thurston 2002, Thronson and Quigg 2008). Fish kills related to hypoxia have been noted in waterbodies on all US coasts, but the incidence of fish kills does not capture the extent of hypoxia impacts. Fish kills are much less widespread than hypoxia itself and sometimes occur only when several factors are present along with hypoxia (e.g., HABs as in Corsica River, Maryland, Bricker et al. 2007) or when hypoxia develops in a way that prevents fish from escaping. Other effects of hypoxia are less obvious, but more pervasive and likely more important overall.

These include shifts in spatial distribution of organisms, changes in community structure caused by emigration of fish and mobile invertebrates, mortality of sessile (immobile) bottom-dwelling organisms, and alteration or blockage of normal migration routes of fish and invertebrates.

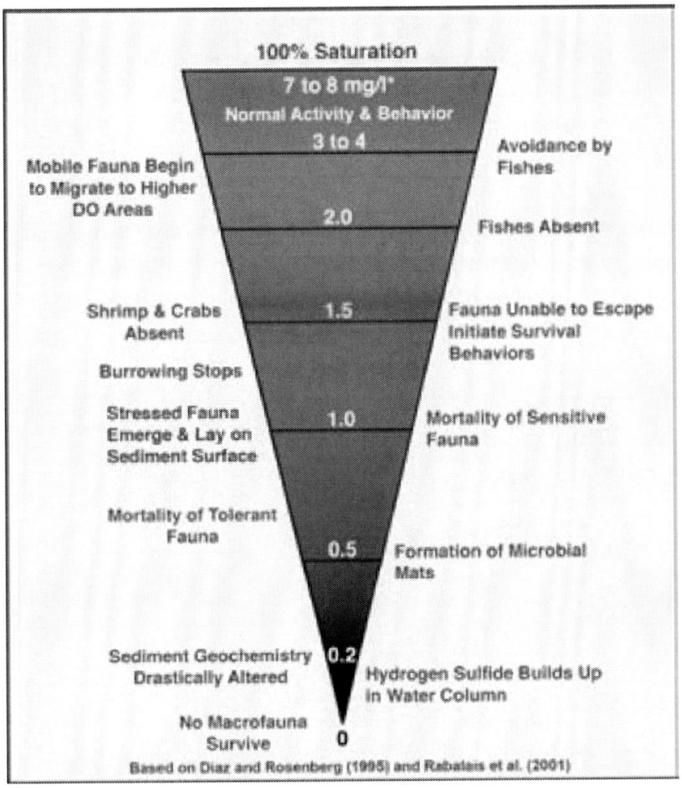

Figure 5. The range of ecological impacts exhibited as dissolved oxygen levels drop from saturation to anoxia (based on Diaz and Rosenberg 1995, Rabalais et al. 2001).

Hypoxic habitats that are avoided by organisms are functionally lost from the system (Breitburg 2002, Rabalais and Turner 2001, Sagasti et al. 2001). Because oxygen thresholds for avoidance are usually higher than for survival (Figure 5), habitat loss due to hypoxia is far greater than would be suggested by calculations based on recruitment or survival tolerances. Changes in community structure and habitat loss associated with hypoxia can also propagate to other components of the food web. For example, impacts to animals that act as crucial food web linkages between algal producers and top predators (e.g., benthic suspension feeders) result in a decreased flow of energy to predators and an increase to microbes (Figure 6, Diaz and Rosenberg 2008). Avoidance of hypoxia can also reduce energy flow to predators. Hypoxia can cause reduced growth (Taylor and Miller 2001), lower reproduction (Marcus et al. 2004), and other physiological effects (Wu et al. 2003). Shrimp and crabs exposed to hypoxia, for example, become immunocompromised and may suffer increased susceptibility to disease and mortality from bacterial infections (Le Moullac et al. 1998, Holman et al. 2004, Burgents et al. 2005; Tanner et al. 2006).

Hypoxia has contributed to the collapse or impairment of a number of commercially important fisheries worldwide, including Norway lobster in the Kattegat Sea (Baden et al. 1990) and bottom-dwelling fish in the Baltic and Black Seas (Breitburg et al. 2001, Mee 2006). Effectively separating impacts of hypoxia on exploited fishery populations from those due to overfishing can be challenging (Breitburg et al. 2009). Most likely, the effect of hypoxia is to decrease productivity and resilience of exploited populations, making them more vulnerable to collapse in the face of heavy fishing pressure.

The ecological impacts of hypoxia may be understood in terms of the ecosystem services normally provided by a healthy ecosystem, but lost as a result of hypoxia (Table 2). A full assessment of ecosystem services lost helps bridge the gap between ecological functions lost and their impact on people. In some cases, though not without challenges, ecosystem services can be assigned a reasonable dollar value. In these cases, analysis of services helps convey the economic costs associated with ecological impacts.

Fisheries yield is one ecosystem service that can be impacted both directly and indirectly by hypoxia. Mortality of fisheries species is a direct mechanism by which services are lost.

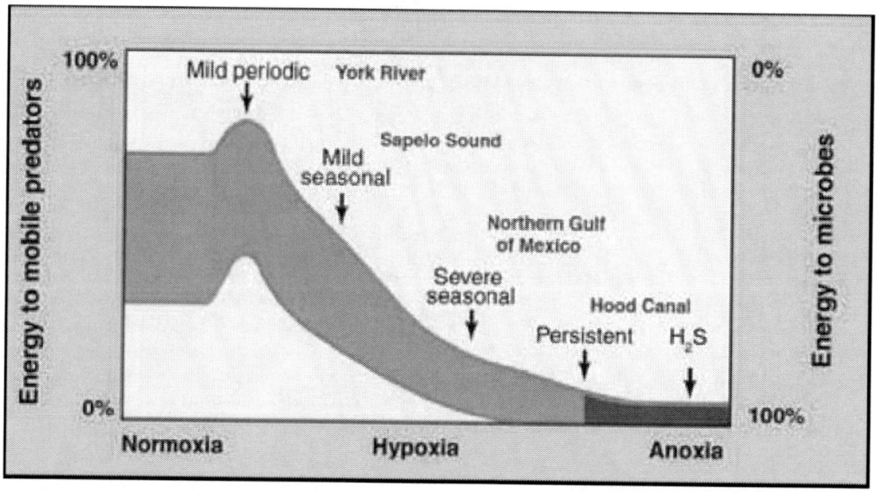

Figure 6. Conceptual view of how hypoxia alters ecosystem energy flow with example systems (modified from Diaz and Rosenberg 2008).

Loss of forage for bottom-feeding fish and shellfish due to hypoxia is probably more important in most cases and also amounts to a loss of ecosystem services. In the Chesapeake Bay, seasonal hypoxia lasts about three months and reduces the Bay's total benthic secondary production by about 5% (Diaz and Schaffner 1990), or roughly 75,000 metric tons of biomass (Diaz and Rosenberg 2008).

This is enough to feed about half the annual blue crab catch for a year. In the northern Gulf of Mexico, severe seasonal hypoxia can last up to six months and reduces benthic biomass by about 212,000 metric tons when the hypoxic zone is 20,000 km^2 (Rabalais et al. 2001).

This lost biomass could feed about 75% of the brown shrimp catch for a season (Diaz and Rosenberg 2008). It is unknown whether these systems can recover secondary production lost to hypoxia during periods with normal oxygen levels.

The Chesapeake Bay has a recovery period of approximately nine months, whereas the northern Gulf of Mexico requires six months for recovery of the lost benthic production. Well-oxygenated conditions outside of summer, however, may be inadequate to support recovery since many other growth requirements (especially appropriate water temperatures) are optimal in the summer.

One of the less obvious ecosystem services lost during hypoxia is sediment mixing by benthic organisms, or 'bioturbation.' Reworking of sediments via bioturbation promotes oxygenation of sediments, improving habitat for benthic animals and promoting biogeochemical feedback processes that reduce nutrient recycling and limit eutrophication. In particular, bioturbation promotes coupled nitrification and denitrification, which eliminates excess bioavailable nitrogen from the ecosystem. Loss of these services amounts to additional nitrogen loading, which must be offset by additional controls on inputs (Hagy et al. 2004). When bioturbation is eliminated due to hypoxia, nitrogen is returned to the water as bioavailable ammonium, often with phosphorus as well, reinforcing a cycle of eutrophication (Aller 1994, Kemp et al. 2005).

Determining the ecological consequences of hypoxia on ecosystem services (Table 2) is often complicated by cumulative and interacting impacts involving a variety of stressors. Quantification of the relationship between hypoxia and impacts on commercial and recreational living resources are needed by managers in order to improve coastal management and policy decision-making. Recent research advances in this area are discussed in Section 3.2.

Table 2. Principal ecosystem characteristics and services impacted by hypoxia. Ecosystem services are processes by which the environment produces resources that are important to humans. Lost ecosystem characteristics highlighted here are critical for supporting ecosystem services. Not listed here are impacts to aesthetics or services lost due to eutrophication in general

Impacts on Ecosystem Services	Example	Reference
Loss of biomass		
Direct mortality of fisheries species Direct mortality of prey species Reduced growth and production Reduced recruitment	Long Bay Northern Gulf of Mexico Chesapeake Bay Patuxent River	Koepfler et al. 2007 Rabalais et al. 2001 Diaz and Schaffner 1990 Breitburg 1992
Loss of biodiversity		
Elimination of sensitive species Reduced diversity Increased susceptibility to disease and other stressors Lower food web complexity	Pensacola Bay Pamlico River St. Johns River Neuse River Estuary	Hagy et al. 2006 Tenore 1972 Mason 1998 Baird et al. 2004
Loss of habitat		
Crowding of organisms into suboptimal habitats Increased predation risks (both natural and fishing) Forced migration from preferred habitat Altered or blocked migration routes	Neuse River Estuary Northern Gulf of Mexico St. Josephs Bay New York Bight	Eby et al. 2005 Craig and Crowder 2005 Leonard and McClintock 1999 Sindermann and Swanson 1980
Altered energy and biogeochemical cycling		
Increased energy flow through microbes Production of toxic hydrogen sulfide Release of phosphorus from sediments Release of ammonia and ammonium from sediments Loss of denitrification	Corpus Christi Bay Dead end canals Lake Erie Chesapeake Bay Chesapeake Bay	Montagna and Ritter 2006 Luther et al. 2004 Hawley et al. 2006 Cowan and Boynton 1996 Lewis et al. 2007

2.4.2. Economic Consequences

There is a growing collection of literature on the ecological consequences of hypoxia, but economic evaluations are lacking. Economic effects attributable to hypoxia are subtle and difficult to quantify even when mass mortality events occur. Much of the problem is related to multiple stressors (habitat degradation, overfishing, HABs, and pollution) acting on targeted commercial populations as well as factors that impact fishers' economics (aquaculture, imports, economic costs of fishing, and fisheries regulations). Economic impacts that stem from the effects of hypoxia on fishery stocks are mostly subtle and tied to ecological impacts through reduced growth and reproduction. Other economic costs imposed on fishers are related to increased time on fishing grounds and costs of searching for stocks (e.g., to reach more distant fishing grounds beyond areas impacted by hypoxia). How these costs translate to impact on profits is complex, however, because in addition to the ramifications of reduced quantity, the unit value of landings on the market affects its total value and must be considered when evaluating the economic impacts.

Although quantifying costs of hypoxia-related mortality events is difficult, there are some published examples (Table 3). Hypoxia in the early 1970s in Mobile Bay, Alabama was estimated to have killed over $500,000 worth of oysters (May 1973). An even greater economic cost was associated with the declining stock size associated with mortality and poor recruitment of oysters in years with severe hypoxia.

Table 3. Examples of hypoxia-related economic impacts. For a current perspective, impacts are also shown in 2009 dollars, calculated from the original values using the consumer price inflation index (http://www.bls.gov/cpi/)

Event	Impacts considered in Estimate	Estimated Economic Impact	Economic Impact in 2009 dollars
Hypoxia in Mobile Bay in 1970's (May 1973)	Mortality to oysters	$500,000	$2,400,000
Modeled hypoxia in Patuxent River, Maryland (Lipton and Hicks 2003)	Striped bass fishing and associated activities in Patuxent River*	$100,000** for dissolved oxygen below 5 mg/L in the River	$110,000**
		$300,000** for anoxia in the River	$340,000**
		$145,000,000** for dissolved oxygen less than 3mg/L in entire Chesapeake Bay (eliminates substitute fishing sites)	$166,000,000**
New York Bight hypoxic event in the summer of 1976 (Figley et al. 1979)	Surf clams, finfishes, ocean quahogs, sea scallops, lobsters	$70,000,000	$265,000,000

* Note that the impact would be substantially greater if all target species were considered.

** Net present value; the net present value represents the chronic effects of hypoxia on the value of striped bass fishing.

A modeling study in the Patuxent River in Maryland estimated that the net value of striped bass fishing alone would decrease over the long-term by over $145 million if the entire Chesapeake Bay were impacted by hypoxia, which would preclude fishing in other sites (Lipton and Hicks 2003).

Impacts of hypoxia on the overall health of the striped bass population and impacts to other Chesapeake fisheries were not included in this estimate but would substantially increase the overall economic consequences to fishers in the region. During the 1976 summer hypoxic event in the New York Bight, acute economic losses were estimated to have been over $70 million (Figley et al. 1979). Much of this loss was due to impacts to the surf clam resource, accounting for more than $60 million. These values consider short-term impacts from actual losses reported by fishers and potential losses due to resource mortality.

Experience with hypoxic zones around the globe shows that both ecological and fisheries impacts become progressively more severe as hypoxia worsens (Diaz and Rosenberg 1995, Caddy 1993) and the area of suitable habitat declines (Bricker et al. 2006). Large systems around the globe have suffered serious ecological and economic consequences from seasonal hypoxia; most notably the Kattegat and the Black Sea. Consequences range from localized loss of catch and recruitment failure to complete system-wide loss of fishery species (Karlsen et al. 2002, Mee 1992). In systems where habitat is only partially lost to hypoxia, such as Long Island Sound and the Chesapeake Bay, catch rates for recreational fishing decline (Bricker et al. 2006).

2.5. Future Considerations

2.5.1. Climate Change

Climate change will almost certainly influence both naturally occurring and eutrophication-related hypoxia, as well as the incidence of other ecological problems, such as HABs. To fully understand environmental change, it is necessary to consider the influence of multiple climate drivers (Table 4). In general, the expected long-term ecological changes favor progressively earlier onset of hypoxia each year and, possibly, longer overall duration (Boesch et al. 2007).

Increasing average water temperature is one mechanism by which climate change may increase susceptibility of systems to hypoxia. Higher water temperatures promote increased water column stratification, decreased solubility of oxygen, and increased metabolic rates, including oxygen consumption and nutrient recycling. Climate predictions also suggest large changes in precipitation patterns, but with significant uncertainty regarding what changes will occur in any given watershed (Christensen et al. 2007). Increased precipitation can be expected to promote increased runoff of nutrients to estuarine and coastal ecosystems and increased water column stratification within these systems, contributing to more severe oxygen depletion (Table 4, Global Climate Change Impacts in the US 2009, Justic′ et al. 2007, Najjar et al. 2000). Climate predictions for the Mississippi River basin suggest a 20% increase in river discharge (Miller and Russell 1992), which is expected to increase the average extent of hypoxia on the northern Gulf of Mexico shelf (Greene et al. 2009). Climate-associated changes in oceanic wind patterns can impact ocean circulation at a large scale. The severe hypoxia that has developed every year since 2002 along the coast of Oregon has been linked to climate-based changes in regional wind patterns, which affect water column stratification and delivery of nutrients from deep water to relatively shallow coastal areas.

Table 4. Estimated influence of climate drivers on the extent and severity of hypoxia (+ = more hypoxia) (Diaz 2008, Modified from Boesch et al. 2007)

Climate Driver	Direct Effect	Secondary Effect	Influence on Hypoxia
Increased temperature	More evaporation	Decreased stream flow	-
		Land-use and cover changes	+/-
	Less snow cover	More nitrogen retention	-
	Warmer water	Stronger stratification	+
		Higher metabolic rates	+
More precipitation	More stream flow	Stronger stratification	+
		More nutrient loading	+
	More extreme rainfall	Greater erosion of soil phosphorus	+
Less precipitation	Less stream flow	Weaker stratification	-
		Less nutrient loading	-
Higher sea level	Greater depth	Stronger stratification	+
		Greater bottom water volume	-
		Less hydraulic mixing	+
	Less tidal marsh	Diminished nutrient trapping	+
Summer winds and storms	Weaker, less water column mixing	More persistent stratification	+
	Stronger, more water column mixing	Less persistent stratification	-
	Shifting wind patterns	Weaker/stronger upwelling potential	+/-

2.5.2. Impacts of Biofuels Strategy

The EISA of 2007 mandates production of 36 billion gallons of biofuels by the year 2022, with a 15 billion gallon limit for the amount to come from corn. Compared to other potential sources of biomass for biofuel production, such as perennial grasses or soybeans, corn is less efficient at taking up applied nutrients and usually requires higher nutrient application rates—these factors contribute to higher nutrient loadings (especially nitrogen) with corn (CBC 2007, NRC 2007). Increased production of corn-based ethanol biofuel is projected to exacerbate hypoxia in the Gulf of Mexico and other coastal areas (Greene et al. 2009, Rabalais et al. 2009). Donner and Kucharik (2008) estimate that expanded corn cultivation will increase the average annual flux of dissolved inorganic nitrogen to the Gulf of Mexico by 10- 34%.

The acreage planted in corn grew by 19% between 2006 and 2007, replacing acreage in conservation reserve programs as well as soybeans and cotton (crops that are frequently less nitrogen-intensive, particularly in the Upper Midwest; see Potter et al. 2006) (Turner et al. 2008, NASS 2007). However, this increase did not continue as high fertilizer prices, favorable prices for other crops, and a return to normal crop rotations led to a 7% decline in corn production in 2008 (NASS 2008). The EISA mandates that the US biofuels strategy be 'sustainable.'

2.5.3. Impacts of Future Management Decisions

The extent of hypoxia in the future will depend on land management practices including agricultural practices.. Climate change will affect water column stratification, organic matter production, nutrient discharges, and rates of oxygen consumption. Land management will affect nutrient budgets and concentrations of nutrients applied to land through agriculture (Figure 7). For example, the expansion of agriculture for production of crops to be used for food and biofuels will, lacking preventive actions, result in increased nutrient loading to coastal waters and increased eutrophication effects (US EPA 2007, Rabalais et al. 2007).

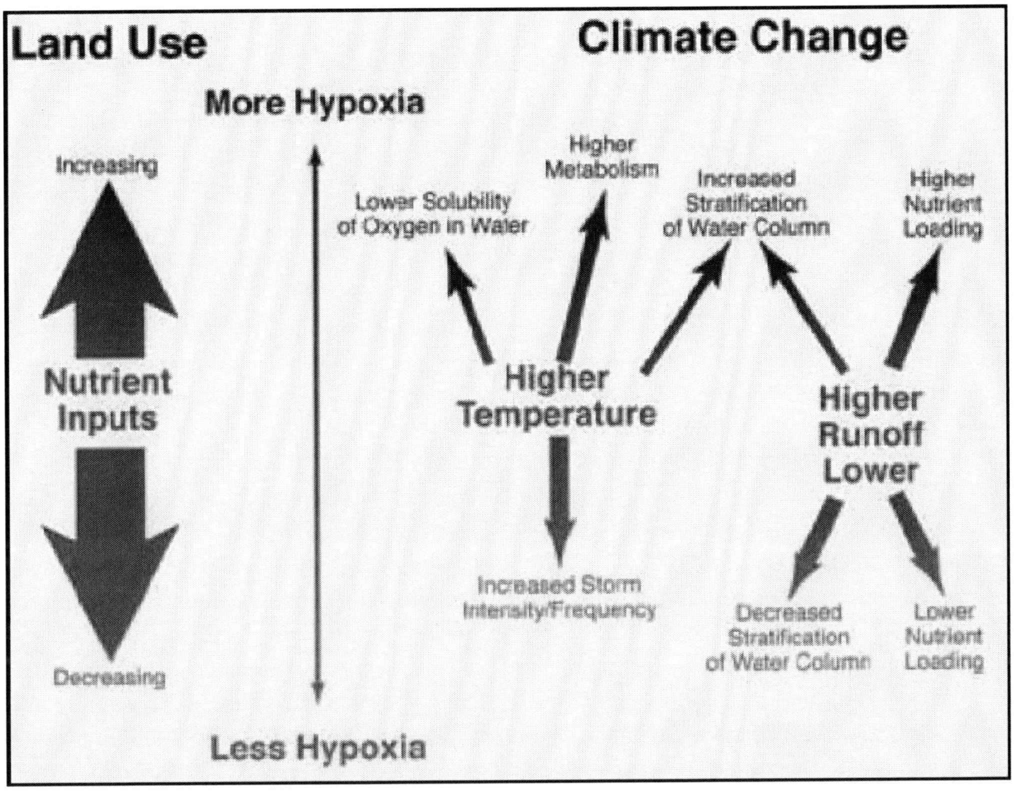

Figure 7. Relative magnitude and contribution (the larger the arrow, the larger the contribution) of land management practices versus climate change factors to expansion or contraction of low dissolved oxygen (modified from Diaz and Breitburg 2009).

On the other hand, improved agricultural conservation practices and large-scale implementation of nutrient best management practices (BMPs) can create dramatically better outcomes for water quality. The development of cellulosic ethanol production will enable farmers and forest managers to derive income from their lands while reducing nutrient runoff. This is an important aspect of the plans to achieve water quality improvements in the Chesapeake Bay region (Chesapeake Bay Commission and Commonwealth of Pennsylvania 2008). Application of these alternatives on agricultural lands that contribute disproportionately high nitrogen loads (e.g., tile-drained fields) will be especially beneficial. More widespread and aggressive implementation of nutrient removal technologies in wastewater and stormwater management also offers opportunities to reduce nutrient enrichment, eutrophication, and hypoxia.

Together, comprehensive nutrient management has the potential to offset the impact of human population pressure, which otherwise will likely continue to be the main driving factor in the persistence and spreading of coastal hypoxia.

3. Federal Hypoxia and Watershed Science Research: Status and Accomplishments

The serious and complex environmental issue of hypoxia and eutrophication cannot be studied or managed by any one Federal or state agency, but requires an organized, comprehensive approach.

This chapter highlights Federal research accomplishments since 2003 when the last HABHRCA-mandated hypoxia assessment was compiled (CENR 2003).

The Federal research presented herein is integral for informed decision-making and successful management of nutrients and hypoxia in an "adaptive management framework" (Figure 8).

Under this framework, current scientific knowledge guides development of management goals with the option to continue to assess knowledge gaps.

As scientific understanding advances, management goals are reassessed and adjusted as needed. Approaches taken in the Gulf of Mexico (Box 4), Chesapeake Bay (Box 5), and Long Island Sound (Box 6) provide good examples of the use of science to inform management decisions.

In addition, regional governance structures involving multiple state and Federal partners are emerging as an important mechanism for mitigating hypoxia because most cases of coastal hypoxia are linked to nutrient loads resulting from watersheds that cross state boundaries. Examples include the Gulf of Mexico/ Mississippi River Watershed Nutrient Task Force, Gulf of Mexico Alliance, Gulf of Mexico Program, Great Lakes Regional Collaboration, Chesapeake Bay Program, and Puget Sound Partnership.

Many of the case studies presented in Appendix II provide additional descriptions of regional governance structures. Given the exponential increase in the number of systems experiencing hypoxia (Diaz and Rosenberg 2008), it is critical that more states take actions to reduce nutrient transport into the coastal zone.

As highlighted in this chapter, each of the various research efforts conducted by Federal agencies or funded extramurally by Federal agencies has focused on understanding a different aspect of the hypoxia issue (Figure 9).

Studies have addressed approaches for reducing nutrients at their sources, what happens to nutrients during transit through the watershed, how nutrients affect hypoxia and related water quality concerns in the coastal zone, and, finally, how hypoxia impacts aquatic life and ecological services provided by coastal systems.

Recipients of extramural funds have included state governments and academic organizations that have coordinated their research with Federal research programs (see Appendix I).

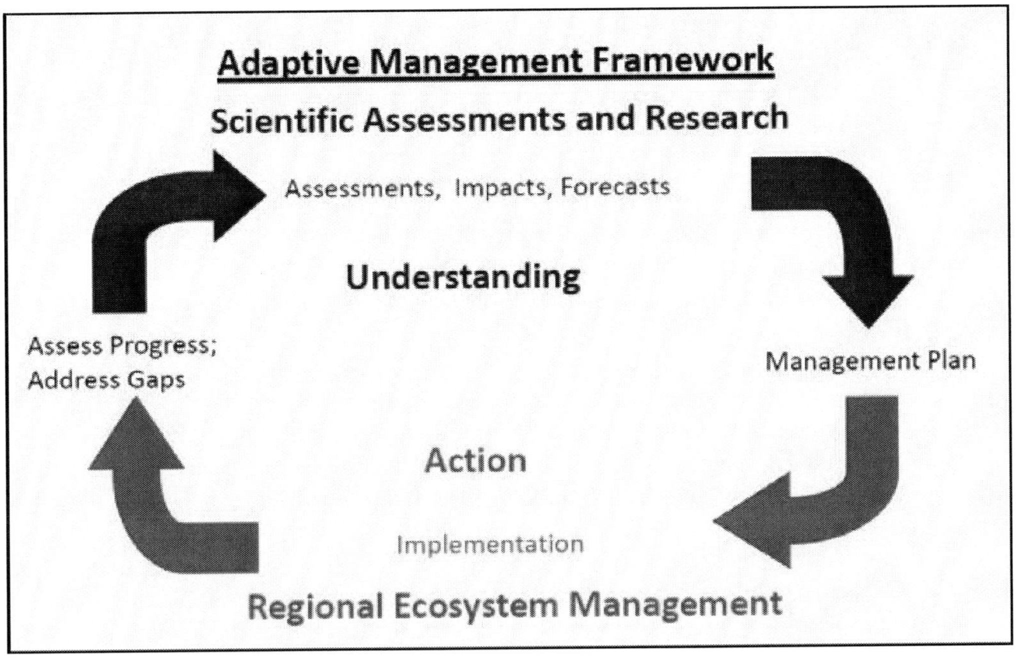

Figure 8. Conceptual diagram explaining how, in an adaptive management framework, scientific research informs management of environmental problems such as hypoxia (and vice versa).

Many Federal agencies contribute to research and management of hypoxia and nutrients. The primary agencies include NOAA, EPA, USGS, and USDA. NOAA has focused research funds and internal capabilities on monitoring and improving understanding of hypoxia and its impacts on commercially- and ecologically-important living resources in coastal waters. EPA bridges the continuum from freshwater ecosystems to estuaries and coastal waters and has focused resources on understanding and regulating nutrient inputs, such as those from wastewater treatment plants, as well as studying and modeling hypoxia in coastal waters. USGS provides critical data through the measurement and modeling of freshwater and nutrient delivery to coastal waters throughout the Nation. Finally, USDA is responsible for developing and implementing strategies to reduce nutrient inputs to coastal waters from agricultural lands and urban ecosystems, which is a major cause of eutrophication and hypoxia in many systems. Recent research accomplishments for these Federal agencies are highlighted in this chapter and in the place-based case studies in Appendix II. Appendix I provides details about programs in Federal agencies which address various aspects of hypoxia. Where appropriate, the Federal agency responsible for conducting or funding specific research accomplishments is parenthetically highlighted in blue text.

3.1. Improving Understanding of Hypoxia in Coastal Waters

Identifying and characterizing the causes of hypoxia are critical for effective ecosystem management and depend heavily upon monitoring programs and the development of models. Differences in factors that cause hypoxia among systems, ranging from primarily nonpoint source nutrients in the northern Gulf of Mexico to wastewater treatment plant effluent in

Narragansett Bay (Figure 3), highlight one source of complexity in mitigating and controlling hypoxia in different coastal systems. The following sections highlight recent monitoring (Section 3.1.1) and modeling (Section 3.1.2) activities and advances in understanding of hypoxia characteristics and causes (Section 3.1.3).

Box 4. Adaptive Management Approach for Gulf of Mexico Hypoxic Zone

HABHRCA 1998 mandated an integrated scientific assessment of hypoxia in the northern Gulf of Mexico. The findings on how and why hypoxia forms in the Gulf informed the first Action Plan signed by Federal and state agencies in 2001. The final recommendation of the 2001 Action Plan called for an updated scientific assessment to be completed in five years which in turn would lead to an updated Action Plan. Following the release of the 2001 Action Plan, the Task Force completed a series of reports, including *A Science Strategy to Support Management Decisions* in 2004 and *The Management Action Review* in 2006. A series of four state-of-science symposia were also completed in 2006. These symposia and reports fed into a comprehensive EPA Science Advisory Board Report. The findings from the reassessment directly informed the 2008 Action Plan, which was signed in June 2008.

The 2008 Action Plan recommends at least a 45% reduction in nitrogen and phosphorus flowing into the Gulf to reduce the size of the hypoxic zone to the 5000 km^2 target. Unfortunately, despite the positive intentions embodied in these various rounds of assessments and plans, the aerial extent of hypoxia in the Gulf has continued to grow, which indicates that an even broader, more forceful, perhaps regulatory, approach is needed.

3.1.1. Monitoring Hypoxia

Monitoring of dissolved oxygen in coastal waters is usually conducted as a component of research and water quality monitoring programs. Many such programs are conducted through partnerships involving one or more agencies of the Federal government, one or more states, and local government. EPA's National Estuary Program (NEP) supports such partnerships around the country in order to implement environmental monitoring. As part of their respective missions, EPA and USGS conduct long-term assessments of environmental and ecological conditions, including dissolved oxygen, within selected bays, estuaries, and other coastal waters. The Integrated Ocean Observing System (IOOS; NOAA), which seeks to integrate all ocean and coastal data within a national monitoring framework, does not currently include dissolved oxygen as one of its five high priority variables for synthesis. However, some of the regional ocean observing systems that contribute to IOOS collect dissolved oxygen data (Appendix I).

There is regionally focused monitoring of dissolved oxygen in the Chesapeake Bay (EPA, NOAA, USGS), Narragansett Bay (NOAA, EPA), Hood Canal, Lake Erie (NOAA), Long Island (EPA), and along the Oregon continental shelf (NOAA, National Science Foundation or NSF; see Figure 10 for examples). Most of the 27 National Estuarine Research Reserves (NOAA–state partnerships, recently summarized by Sanger et al. 2002) monitor dissolved oxygen through the System-Wide Monitoring Program (SWMP).

Finally, routine monitoring of dissolved oxygen as part of juvenile salmonid surveys from Oregon to the Canadian border along the shelf began in 2006 in order to monitor the upwelling dynamics that have caused periodic hypoxia off the coast of Oregon (NOAA).

Monitoring in the northern Gulf of Mexico, the largest hypoxic zone in the United States (Figure 10d) and one of two hypoxic zones located in Federal waters (the other one is off the Oregon coast, see Appendix II) has received increasing attention in recent years. Since 1985, an annual shelf-wide survey has been conducted during July, when hypoxia is presumed to be most widespread. Although this has been funded as part of a competitive research program, the surveys have provided the principal metric for assessing progress toward the goal of a reduction in hypoxia outlined by the Gulf of Mexico/Mississippi River Watershed Nutrient Task Force (see Gulf of Mexico case study, Appendix II) and for validating hypoxia forecasts (NOAA). Other ship-based sampling of water quality (including dissolved oxygen) on the continental shelf and near major outlets of the Mississippi and Atchafalaya Rivers occurs annually and seasonally as a part of shelf-wide surveys and hypoxia-related research (NOAA, EPA). These surveys were designed to quantify key physical and biogeochemical processes influencing the development and persistence of hypoxia, and to support development of predictive models (NOAA, EPA).

Box 5. Adaptive Management Approach for Chesapeake Bay

Chesapeake Bay Program
A Watershed Partnership

In 1983, following the conclusion of a 5-year EPA Chesapeake Bay Study and other water quality initiatives, Bay states and the Federal government agreed to reduce nitrogen inputs into the Bay through the first Chesapeake Bay Agreement. A Scientific and Technical Advisory Committee was formed at this time and released its first report in 1986. The scientific information and consensus in this report led to the Second *Chesapeake Bay Agreement* in 1987, calling for a basin-wide nutrient reduction strategy and more stringent nutrient reduction targets. Reevaluations in 1992 and 1997 then led to the *Chesapeake 2000 Agreement*, which recommitted to goals of the prior agreements while outlining more specific strategies for achieving the targets. In support of the adaptive management process, the Chesapeake Bay Program recently agreed to a framework for a five-stage adaptive management model (Kaplan and Norton 2008).

A recent assessment of water quality trends in rivers feeding the Bay showed significant improvements for nutrient loading for nitrogen (72%), total phosphorus (81%), and sediment (43%), indicating that management actions are having some effect in reducing nutrients and sediments (Langland et al. 2006). However, to date, the management of nutrient input in the Chesapeake Bay has not improved dissolved oxygen, although it has caused small-scale reversals in hypoxia (Diaz and Rosenburg 2008).

Seasonal field surveys using EPA's offshore survey vessel and satellite ocean color remote sensing were conducted in 2003, 2006, and 2007 to characterize the magnitude and variability in physical, biological, and chemical oceanographic conditions and processes along the northern Gulf continental shelf from the Mississippi River Delta to Texas (EPA). NOAA's Gulf of Mexico Hypoxia Watch (http://ecowatch.ncddc. noaa.gov/hypoxia) has provided internet access to summer seasonal dissolved oxygen data collected through routine the Southeast Area Monitoring and Assessment Program (SEAMAP) groundfish monitoring cruises in June and July (NOAA). Several moored instrument arrays have also been deployed in the Gulf of Mexico hypoxic zone to obtain continuous records of dissolved oxygen. Efforts to improve and test the moored instruments are continuing (NOAA, Minerals Management Service).

Recognizing the need to expand coastal monitoring in the northern Gulf of Mexico and move funding from a competitive research mechanism to an operational monitoring framework, NOAA convened a workshop in January 2007 entitled "Summit on Long-Term Monitoring of the Gulf of Mexico Hypoxic Zone: Developing the Implementation Plan for an Operational Observation System." Three committees were formed as a result of the Summit executive, technical, and stakeholder) and have worked to develop a monitoring implementation plan, which includes detailed designs and cost estimates. This plan was completed in December 2008. Operational monitoring for the northern Gulf of Mexico is important for tracking progress in remediating hypoxia and support forecasts and other scientific research needed to inform management. Implementation of this plan is a priority.

Box 6. Sound Science Leads to Significant Reductions in Hypoxia in Long Island Sound

Following progressive declines in water quality and increasing prevalence of hypoxia, the Long Island Sound Study (LISS) was formed in 1985 to guide the recovery of the Sound. The LISS released a Comprehensive Conservation Management Plan in 1994 that outlined 3-phases: 1) in 1990, freezing nitrogen loadings in critical areas to prevent worsening of hypoxia, 2) implementation of low-cost nitrogen reduction measures, and 3) reduction of nitrogen loads to target levels.

In 1998, the LISS adopted a 58% nitrogen reduction target by 2014 and in 2001, the EPA approved state total maximum daily loads (TMDL) plans to achieve this goal. By 2005, point source nitrogen loads had been reduced by 20% from the peak in the early 1990's and the severity of hypoxia declined (see Long Island Case Study in Appendix II).

LONG ISLAND SOUND STUDY

Box 7. Hypoxia Advanced Warning Protects Drinking Water

Since 2006, when three out of four Cleveland water treatment plans were exposed to anoxic waters carrying increased loads of certain metals, NOAA deployed water quality monitoring sensors as part of its Real-time Coastal Observation Network (ReCON) near the plants' water intakes. The ReCON sensors provided advanced warning of hypoxia in 2007, allowing the Cleveland Water Division to adjust their water treatment approach to maintain the safety of water supplies for 1.5 million people.

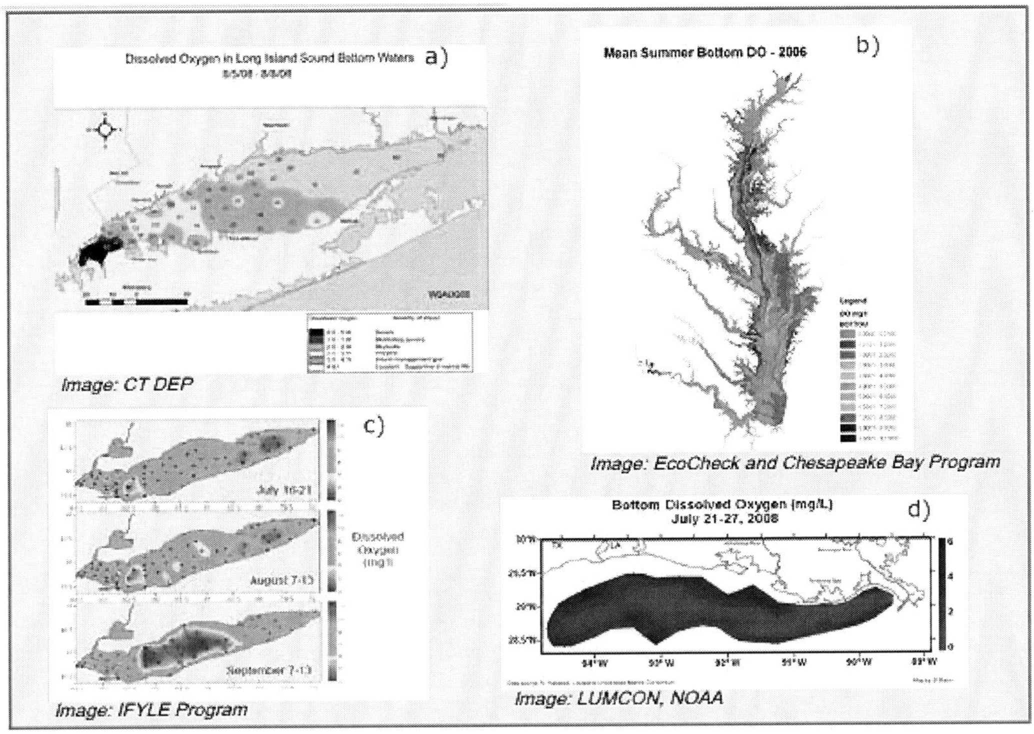

Figure 10. Hypoxia is most intensively monitored in the largest and most impacted coastal systems in the United States. Examples include: a) Long Island Sound, b) Chesapeake Bay, c) Lake Erie, and d) Northern Gulf of Mexico.

3.1.2. Modeling and Forecasting Hypoxia

Environmental models can range from simple statistical equations to complex, computer-intensive three-dimensional simulations that incorporate various aspects of physical, chemical, and biological processes in a coastal waterbody and its watershed.

Models are essential tools for making scientific inferences when systems are too large or complex to conduct manipulative experiments. Weather forecast models are the most widely appreciated environmental models, but environmental models are also widely used for air quality forecasting and management, fisheries management, emergency management, and water quality management, among others. Models are particularly useful for understanding hypoxia because hypoxia reflects strongly coupled physical-chemical-biological processes. Computer intensive water quality models have been used for nearly three decades to predict the impact of nutrient management actions on specific water quality outcomes, including the extent and severity of hypoxia.

Table 5. Models developed since 2002 to forecast or simulate Gulf of Mexico hypoxia (adapted from Justic´ et al. 2007)

Model	Class	Key Outputs	Key Attributes
Greene et al. (2009)	Statistical	Substantial and sustained nitrogen and phosphorus reductions are required to reduce hypoxia.	Indicated dependence of conclusions on model choice and input data and provides estimates of forecast uncertainty.
Justic´ et al. (2002)	Simulation	Riverine nutrients play a major role in the development of hypoxia.	Provides estimate of time series of oxygen at a single location. Includes mechanistic detail.
Scavia et al. (2003, 2004)	Simulation	Extensive hypoxia not common before mid-1970's. A 40-45% reduction is nitrogen may be required to meet management goals.	Provides estimates of forecast uncertainty using a biophysical modeling scheme.
Stow et al. (2005)	Statistical	Stratification is a predictor of hypoxia and the amount of stratification required to induce hypoxia has gone down (1982-2002).	Uses statistical approach to characterize long-term change in hypoxia relative to stratification strength.
Turner et al. (2005, 2006)	Statistical	Hypoxic zone has only recently developed (1970's or 1980's) and nitrogen is the major driving factor controlling the size.	Incorporates spring nitrate, river discharge, and year factor to predict hypoxia
Hetland and DiMarco (2008)	Simulation	Biological processes controlling hypoxia vary from east to west. Hypoxia is formed in place, and not advected across shelf.	Uses 3-dimensional hydrodynamic model with simplified water quality model to explore implications of different controls on oxygen consumption for the spatial distribution of hypoxia.

Coupled watershed-water quality models have been a critical tool in the effort to manage nutrients, reduce hypoxia, and, more broadly, restore the health of the Chesapeake Bay ecosystem. The EPA Chesapeake Bay Program has funded water quality model development by the US Army Corps of Engineers Waterways Experiment Station and others since 1987 and has relied heavily on the resulting models to guide nutrient reduction strategies throughout the Chesapeake Bay watershed (Koroncai et al. 2003; Cerco and Noel 2004, 2005). Model results provide options to be analyzed in the course of determining a nutrient management strategy. A deliberative process based on model results is necessary because all models, even if very useful, are imperfect and subject to error (Koroncai et al. 2003).

While efforts to improve Chesapeake Bay modeling are ongoing, recent Federal modeling efforts have focused heavily on model development in the northern Gulf of Mexico hypoxic zone, which has followed a somewhat different trajectory. A variety of relatively simple models have been developed to forecast the response of hypoxia to various nutrient load scenarios. In addition, more complex models have been and are being developed to improve understanding of basic scientific questions (Table 5). Coupled hydrodynamic-water quality models have been developed for the northern Gulf but these models lack important mechanistic details, particularly in their representation of biogeochemical and other biological processes (Table 5).

Despite being recently developed and still relatively basic, models have nonetheless been very influential in setting nutrient reduction targets for hypoxia in the northern Gulf of Mexico (US EPA 2007, Justić et al. 2007; NOAA). Simple regression models relating hypoxia to either nutrient loading (Turner et al. 2006b; NOAA) or freshwater flow and nutrient concentration (Greene et al. 2009; EPA) have proven surprisingly predictive. The model by Turner et al. (2006b) has been used annually since it was published to forecast the expected extent of hypoxia in midsummer. A suite of regression models (Greene et al. 2009; EPA) explore how model selection may influence possible conclusions and applications. Both studies suggest that a decrease in nutrient loads from the Mississippi Atchafalaya River Basin (MARB) on the order of 45% will be sufficient to reduce the extent of hypoxia on the shelf to the goal (five-year average of 5,000 km^2) established in the first Gulf Hypoxia Action Plan. Turner et al. (2006b) suggest that hypoxia was not present on the shelf prior to the mid 1970s, whereas Greene et al. (2009) conclude that hypoxia likely occurred as early as the 1950s, but most likely was much less extensive and perhaps did not occur every year as it presently does.

Scavia et al. (2003, 2004; NOAA) employ a biophysical model to parameterize the physical regime on the northern Gulf of Mexico shelf and forecast hypoxia extent (Figure 11). Scavia et al. (2003, 2004) and Greene et al. (2009) each provide models that forecast the extent of hypoxia as a function of nutrient inputs with quantified bounds of forecast uncertainty, a useful development for management applications. These essentially empirical models provide a scientific rationale for the 45% reduction in both nitrogen and phosphorus loading that has been suggested to be needed in order to achieve the hypoxia reduction target for the Gulf. They lack, however, explicit simulation of ecological processes that might be used to better address fundamental questions regarding ecological controls on the extent and intensity of hypoxia as well as the impacts of hypoxia. For these applications, process-oriented dynamic simulation models would be preferable. These models would be more similar to the coupled hydrodynamic-water quality models that have been developed for the Chesapeake Bay (Cerco and Noel 2004, 2005).

For simulation models, the Regional Ocean Model System (ROMS), a multi-purpose, multidisciplinary, open source oceanic modeling tool, is being adapted to Narragansett Bay and will be coupled with an ecological model to improve understanding of the conditions driving hypoxia in that system (NOAA). The ROMS model is also now functional for the Chesapeake Bay, and a new project is focused on adding ecological and sediment components to predict impacts of climate and nutrients on hypoxia and the ecosystem (NOAA). The ROMS ecological hybrid model is also being adapted for the finer-scale resolution existing in the shallower, narrower Delaware Inland Bays which periodically experience fish kills due to severe diel hypoxic events (NOAA). A modified version of the ROMS model uses a number of oceanographic and meteorological variables to examine the effects of water column stratification and circulation on hypoxia in the northern Gulf of Mexico (Hetland and DiMarco 2008, NOAA). Additionally, Justic′ et al. (1996, 2002; NOAA) developed another simulation model, a coupled biological-physical model to examine oxygen cycling dynamics in the core of the hypoxic zone. This model used a number of forcing functions including nitrate loads, monthly Mississippi River runoff, and oceanographic and meteorological conditions. It enabled the simulation of historical oxygen concentrations and the linking of Mississippi River nutrient loadings to eutrophication.

Models can also be used to assess estuarine system susceptibility to hypoxia as a result of nutrient enrichment. An initial version of a Nitrogen-Phytoplankton-Detritus-Oxygen (PDO) model is being developed to address estuarine susceptibility to nutrient enrichment, considering the effects of nutrient loading, flushing, stratification, denitrification, and temperature (NOAA). A preliminary PDO model is being tested on 87 estuaries and shows promise in predicting chlorophyll-*a* concentrations across a range of estuarine types. The model could be used to determine which estuaries, especially ones not currently impacted, are most susceptible to secondary effects of enrichment. This information could help drive preventative management actions.

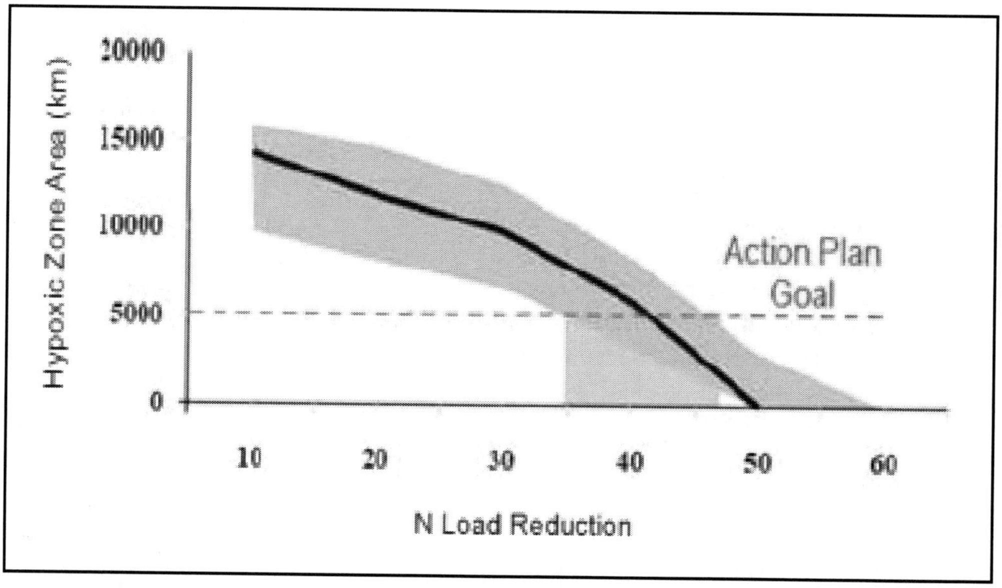

Figure 11. Ensemble forecasts of the response of hypoxia to changes in riverine nitrogen load. Loads are percent reduction from the 1980–1996 mean May–June total nitrogen loads. The grey shaded area, representing the ensemble forecast, contains all values from 1,000 simulations. The horizontal bar at 5,000 km2 represents the Action Plan goal. The blue area represents the range of nitrogen load reductions needed to achieve that goal. Most of the Gulf of Mexico hypoxia models agree that an approximately 45% reduction in both nitrogen and phosphorus are required to reduce the size of the hypoxic zone to the goal set by the Mississippi River/Gulf of Mexico Watershed Nutrient Task Force (adapted from Scavia et al. 2003, 2004).

Finally, as part of the International Field Years on Lake Erie (IFYLE, NOAA), an international government and academic partnership, a major effort has been underway since 2005 to model various aspects of the Lake Erie ecosystem with a special emphasis on understanding the causes and dynamics of hypoxia. In an effort to develop a modeling framework that could help synthesize the data collected during IFYLE, a new project will employ a number of modeling approaches (e.g., watershed, river hydrology, lake hydrodynamics, lower food web, spatially explicit higher food web, statistical) to help identify the causes of hypoxia in Lake Erie and its consequences for the Lake Erie food web (NOAA).

3.1.3. Identifying Causes and Characteristics of Hypoxia

Chapter 2 provides an overview of the general causes of hypoxia. In addition, Appendix II covers a series of specific case studies which provide additional detail about causes and characteristics of hypoxia in highlighted systems. In general, advances have been made in understanding the relationship between various nutrients and the size or extent of hypoxia. For instance, determining that phosphorus is a limiting nutrient for primary production on portions of the continental shelf in northern Gulf of Mexico waters led to a call for reduction in both nitrogen and phosphorus levels being transported down the Mississippi River (NOAA, EPA). In addition, there has been improved understanding of the influences that physical oceanographic variables (e.g., salinity and currents) have on hypoxia formation and characteristics.

Understanding what aspects of the physical system allow hypoxia to develop and tracking how it changes through time have been active areas of research for most systems that experience hypoxia.

Additionally, hypoxic waters tend to remain near where they were formed, suggesting that currents do not move hypoxia across the shelf (NOAA). The structural features (horizontal and vertical characteristics) of hypoxia have been accurately reproduced using a combination of benthic respiration and hydrodynamic models. This has reaffirmed the observed spatial and temporal variability of the hypoxia, especially towards the end of the hypoxia season. Different processes regulate the formation of hypoxia in different regions of the hypoxic zone. Additionally, despite being disrupted by hurricanes (which serve to mix the water column and oxygenate the bottom layer), the hypoxic zone can be resilient, quickly reforming once the storm has passed (NOAA).

The physical circulation of Narragansett Bay has been well studied through the use of strategically deployed shipboard and moored instruments. Recent results show that stratification is a major driver for hypoxia, with hypoxia happening most often in summer several days after a neap tide (NOAA). Meanwhile, efforts to determine the causes of hypoxia off the Oregon shelf (e.g., Chan et al. 2008) are being pursued after hypoxia and anoxia in waters less than 50 m deep caused massive mortality of benthic organisms (NOAA, NSF). Upwelling dynamics are important for many of the west coast systems experiencing hypoxia such as Yaquina Bay and Hood Canal (see Appendix II).

National assessments by a variety of Federal agencies and Federal-state partnerships, with the support of academic scientists, have revealed and characterized widespread nutrient pollution in estuarine waters (CENR 2000, 2003; NRC 2000; US Commission on Ocean Policy 2004).

The EPA Water Quality Inventory reported that 17% of the estuarine waters assessed were impaired by low dissolved oxygen due to anthropogenic eutrophication (US EPA 2002). EPA's NEP and National Coastal Assessment Program and the NOAA's National Centers for Coastal Ocean Science have been formally monitoring and assessing individual estuaries since the early 1990s and have broadly reported that estuarine water quality has been degraded by excess nutrients and eutrophication (US EPA 2005, 2008; Bricker et al. 1999, 2007).

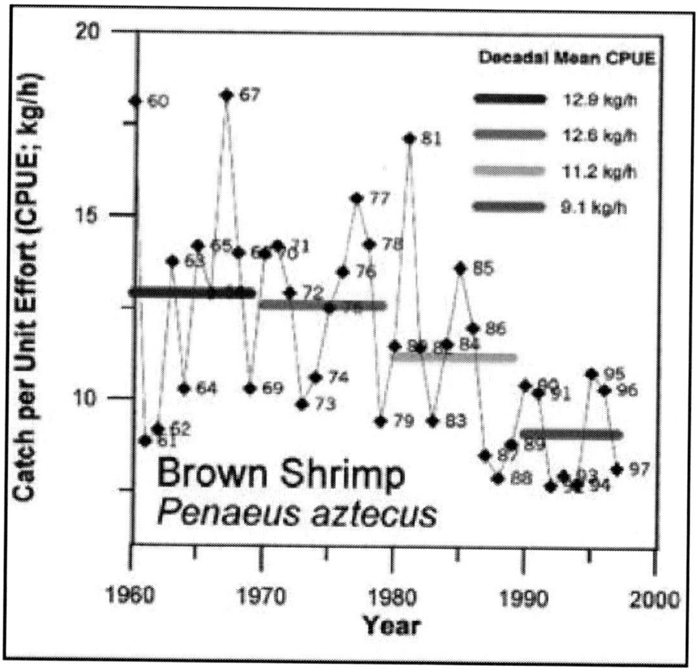

Figure 12. Trends in catch per unit effort for brown shrimp in the northern Gulf of Mexico (modified from Downing et al. 1999).

The first *National Estuarine Eutrophication Assessment* (Bricker et al. 1999, NOAA), a survey of the extent, severity, types, probable causes, and future outlook of eutrophic symptoms, was conducted in the early 1990s. An update, examining the status of eutrophication in the early 2000s and the changes that have occurred since the early 1990s, was completed in 2007 (Bricker et al. 2007). For both reports, data and information were collected for 141 estuaries and coastal waterbodies from scientists and resource managers from academia, state, and Federal agencies (i.e., USGS, EPA, and NOAA), as well as nongovernmental organizations. This national assessment emphasizes the need for a coordinated and integrated effort that balances management action, efficient monitoring to assess the effectiveness of the management, focused research, and a communication campaign aimed at engaging the broader public.

3.2. Quantifying and Modeling the Impacts of Hypoxia

As discussed in Chapter 2, quantifying the impacts of hypoxia remains a critical information gap. The state-of-knowledge related to hypoxia impacts, as well as the research and management needs, were assessed in the NOAA-supported "Ecological Impacts of Hypoxia on Living Resources Workshop," which was held in March 2007. A peer-reviewed special issue of the *Journal of Experimental Marine Biology and Ecology*, which resulted from the 2007 workshop and contains the most up-to-date knowledge of hypoxia impacts, was published in 2009 (Lewitus et al. 2009). The major results from this workshop and other Federally funded research are presented below.

Quantifying the ecological impacts of hypoxia is essential for developing meaningful targets for hypoxia mitigation strategies (e.g., goals for areal extent or periodicity of hypoxia). A number of effects on living resources have been documented in both field and laboratory experiments. Many of these effects are dependent on both the temporal and spatial scale of hypoxic events and can be manifested through both direct and indirect impacts. Current research suggests that effects at both the population and individual levels are likely to be revealed not only through direct means, such as fish kills, but also through complex indirect, often sub-lethal mechanisms, such as reduced reproductive output and growth. However, separating the effects of hypoxia from other stressors that also negatively influence populations and individuals is extremely difficult. For example, trends in the catch per unit effort of brown shrimp in the northern Gulf of Mexico are consistent with impacts from hypoxia, habitat loss, overfishing, and economic factors (Figure 12). But the area of hypoxia may have a greater influence on annual landings (Figure 13). Models which relate experimental field and laboratory studies to population-level changes in fisheries and potential economic impacts are being developed for several systems including the Neuse/Pamlico River estuary, Lake Erie, Chesapeake Bay, Delaware Inland Bays, and northern Gulf of Mexico (NOAA).

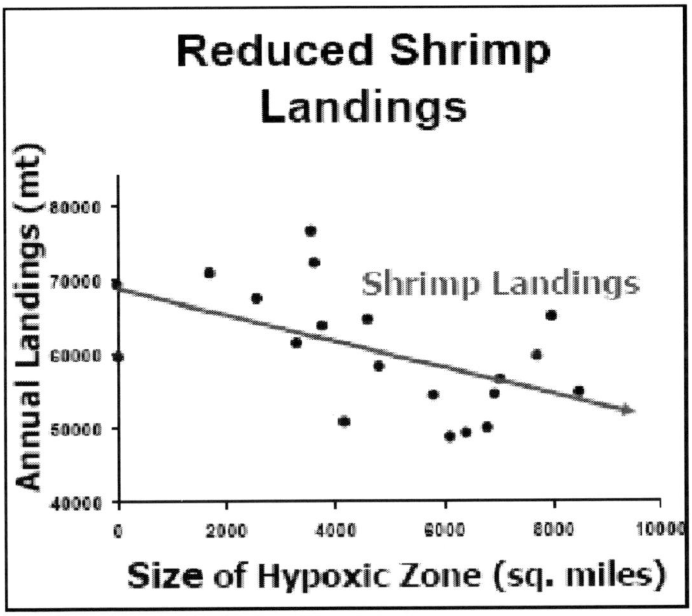

Figure 13. Relationship between annual landings of brown shrimp in the northern Gulf of Mexico and the size of the hypoxic zone (O'Connor and Whitall 2007).

3.2.1. Mortality and Altered Distribution of Fauna

Faunal mortality (e.g., fish kills) has been documented in most coastal systems affected by hypoxia and is often a factor of the duration and extent of hypoxic conditions. Most of the species affected are invertebrates that are unable to move (e.g., clams) when dissolved oxygen becomes too low. This has been best documented in the Chesapeake Bay, where summer hypoxia typically causes mortality among immobile bottom-dwelling organisms (e.g., clams, mussels, and worms) in deep water (see Seitz et al. 2009). Similarly, mortality events in the northern Gulf of Mexico are typically limited to immobile benthos within the hypoxic zone, although mortality of mobile fauna has also been reported (see Rabalais et al. 2001). Elsewhere, the increasing frequency of transient hypoxic events in Puget Sound has resulted in large localized die-offs of fish and shellfish species, while die-offs of menhaden and other species have been increasing in frequency in the Neuse River, North Carolina and in Narragansett Bay, Rhode Island. These areas were also sites of benthic faunal die-offs, especially in Narragansett Bay where hypoxia has resulted in localized loss of blue mussel populations (Altieri and Witman 2006).

Although it has been long known that mobile organisms, especially fish, avoid hypoxic waters, the extent and pattern of movements have only recently been documented. In most systems, including the Chesapeake Bay and the northern Gulf of Mexico, virtually all bottom-dwelling mobile species (e.g., Atlantic croaker, Craig and Crowder 2005), as well as some open-water species (e.g., bay anchovy, Taylor et al. 2007) are absent in bottom waters of hypoxic areas, while in Lake Erie fish are typically not found beneath the thermocline where dissolved oxygen can be low (Ludsin et al. 2009). Within Delaware's coastal bays, juvenile weakfish avoid hypoxic conditions caused by diel oxygen cycling, resulting in the daily emigration of weakfish out of tidal tributaries and their preferred habitats (Tyler and Targett 2007). Movements of Atlantic croaker and shrimp in the Gulf of Mexico have produced a "halo effect" which describes the congregation of fish around the edge of the hypoxic zone (Craig and Crowder 2005, Craig et al. 2005). Changes in zooplankton distribution have also been observed in Lake Erie, the Gulf of Mexico, and the Chesapeake Bay, with changes in biomass size spectra detected in the latter two systems. There have also been anecdotal reports that commercial fishing has adapted to hypoxia-induced altered spatial distribution, possibly placing concentrated prey at risk of overfishing.

As would be expected, an analysis of fishery landings in 22 ecosystems that experience seasonal hypoxia (> 40% of total bottom area affected) determined that pelagic species such as Clupeoid fish were more abundant than bottom-dwelling species such as crustaceans. Systems experiencing increased nutrient loadings, but not hypoxia, experienced the opposite effect on species composition, with more bottom-dwelling crustaceans than pelagic fish (Breitburg et al. 2008).

Hypoxia-induced population migrations likely reflect an effective loss of preferred habitat for some species (Figure 14). The congregation of Atlantic croaker and brown shrimp along the edge of the hypoxic zone in the Gulf of Mexico (the halo-effect) suggests a loss of preferred habitat. Further, white shrimp spend a critical portion of their life cycle within the Gulf of Mexico and can be excluded from optimal spawning habitat. Hypoxia in the Chesapeake Bay has reduced and fragmented habitat for endangered sturgeon, likely hampering the species recovery (Niklitschek and Secor 2005). In addition to exclusion from preferred physical habitat, migrational patterns of species have been altered both in the Gulf of Mexico (e.g., brown and white shrimp) and along the coast of New Jersey (e.g., blue fish, Garlo et al. 1979).

3.2.2. Changes in Food Web Structure and Physiology

An emerging, indirect impact of hypoxia is the alteration in predator-prey relationships and food web functions that often involve complex pathways through cascading effects. Shifts in diets can result in changes in the transfer of energy through the food chain. Hypoxia-induced shifts in diet have been documented in a number of systems and typically result in shifts from bottom-dwelling to pelagic food items (Baird and Ulanowicz 1989, Baird et al. 2004). Additionally, the availability of suitable prey can be limited for planktivorous fish (fish that eat plankton) during severe hypoxia, due to hypoxia-induced changes in plankton distribution and segregation from predators (Zhang et al. 2009). Conversely, it has been hypothesized that the altered spatial distribution of organisms can also increase predation by concentrating prey, such as along the edge of a hypoxic area. There can also be benefits to both prey and predators, but these are species dependent. For example, in Narragansett Bay, quahogs benefited from reduced predation pressure during moderate hypoxic conditions, whereas blue mussels suffered high mortality under similar conditions and thus received no benefit from reduced predation (Altieri 2008). Varying tolerance to hypoxia can also increase the abundance of certain predators. For example, hypoxia can lead to increased relative abundance of gelatinous zooplankton (jellyfish and comb jellies) which are more tolerant to low dissolved oxygen, which in turn leads to increased predation on fish larvae. The overall effects of hypoxia on food web structure have only recently begun to be quantified, but appear to be both species- and system-specific. Sustained, long-term hypoxia can ultimately lead to a total shift in food web structure, which has been observed in several systems (Rabalais et al. 2001, Baird et al. 2004).

Researchers are increasingly examining sub-lethal physiological, reproductive, and bioenergetic impacts of hypoxia. A reported effect of hypoxia in commercially- and recreationally-relevant species is growth reduction. In the Gulf of Mexico, hypoxia has been linked to reductions in brown shrimp lipid content as well as size. In laboratory analyses, striped bass and menhaden growth were reduced, while field examinations have found reduced growth in both juvenile summer flounder and weakfish.

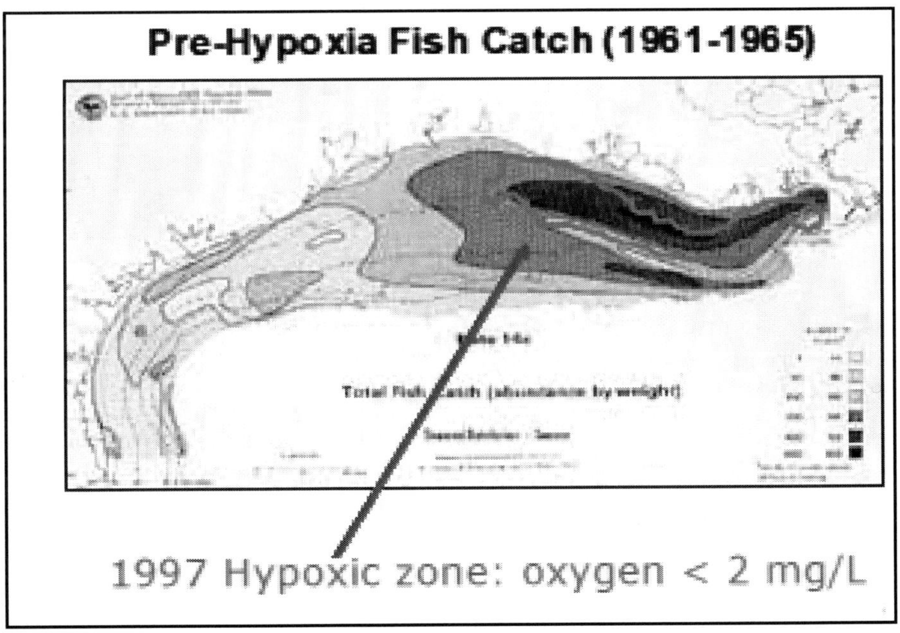

Figure 14. Map of Gulf of Mexico with darker shaded areas indicating denser fish populations in the 1960s. Red line outlines current general configuration and geographic location of the hypoxic zone. Source: Kevin Craig, FSU.

Reproductive effects include alterations to reproductive organs, reduced production of offspring, and lowered rates of offspring survival. Atlantic croaker have also exhibited altered endocrine function and suppressed growth of ovaries and testes under similar conditions. Reduced egg production attributed to the effects of hypoxia has been found in Atlantic croaker, grass shrimp, and naked gobies in the Chesapeake Bay and Gulf of Mexico estuaries (Thomas et al. 2007). Further, egg survival can be reduced when eggs spawned within the water column sink into hypoxic water, as has been observed with bay anchovy and weakfish in the Chesapeake Bay (Breitburg 2002). Finally, a number of other sublethal or indirect physiological effects of hypoxia have also been documented including immune suppression in oysters (Anderson et al. 1998) and reduced consumption and respiration in fish and shellfish.

3.3. Monitoring and Modeling Nutrient Flux in US River Systems

Monitoring streamflow and nutrient fluxes in river systems that feed coastal estuaries is essential to understanding and predicting coastal hypoxia. Furthermore, knowledge of the sources of nutrients ultimately delivered to the coastal zone and knowledge of processes that reduce nutrient loads during transit are required for the development of effective management strategies for reducing coastal hypoxia. It is not economically feasible to monitor all the sources and fluxes of nutrients in all rivers and streams in the United States. However, models using monitoring data from targeted watersheds, representative in terms of their climate, geography, land use, and management measures, can be extremely valuable tools. These models can identify the geographic areas with the highest nutrient contributions, the human activities that contribute the most nutrients, the ecosystems that reduce nutrient fluxes, and potentially the management measures that are most effective at mitigating nutrient fluxes.

3.3.1. Monitoring Nutrient Sources and Fluxes to Coastal Systems

There are two major components of riverine monitoring in the United States. The first component is monitoring delivery of streamflow and nutrient fluxes to coastal environments. This monitoring requires sufficient temporal resolution to enable development of cause and effect linkages between the timing of nutrient fluxes and the development of hypoxia. The second component comprises monitoring of inland watersheds. This monitoring happens in watersheds of various scales and supports modeling applications that define the spatial distribution of the sources of nutrients and the human activities that affect those fluxes.

The USGS is the primary agency monitoring streamflow, nutrients, and other water quality constituents in rivers and streams throughout the United States. Other agencies rely on these data to determine effective conservation measures to protect downstream resources. Additional monitoring is conducted by state governments through water quality compliance programs under EPA oversight. Although limited by the objective and design, more and more of these data are being integrated into regional and national monitoring efforts.

An important element of the design of programs monitoring nutrient delivery to coastal systems is the availability of continuous streamflow records and water quality measurements taken with sufficient frequency to enable accurate estimation of nutrients fluxes (mass per unit time) with adequate temporal resolution. The US Commission on Ocean Policy (2004) cited a critical loss of USGS stream gauging capacity due to funding constraints, which has limited the ability to monitor water and other constituents entering our coasts.

Most monitoring programs are designed to measure the concentration of nutrients at the time of sampling, and measurements are only taken a few times each year. However, flux estimates are essential to evaluating the causal relationship between nutrient delivery and hypoxia and to evaluate trends, including the performance of management actions.

The USGS National Stream Quality Accounting Network was redesigned in 2007 as a national coastal monitoring network with the primary goal of measuring delivery of streamflow and the associated fluxes of nutrients and selected other chemical constituents to coastal environments. The redesign maximizes existing resources and ongoing monitoring programs (such as the USGS National Water Quality Assessment Program and state monitoring programs). In 2008, five additional stations were added to this national coastal monitoring network as a result of the initial steps of implementation of the National Monitoring Network of the Ocean Action Plan, thereby increasing the number of large coastal rivers being monitored from 13 to 18. These 18 stations were selected because the rivers they monitor contribute the greatest percent of streamflow and nutrient fluxes to coastal environments. Seventeen stations are in the conterminous United States; the other is on the Yukon River in Alaska. Together, these stations measure more than 80% of the total discharge of streamflow, nitrogen, phosphorus, and suspended sediment to US coastal waters (USGS). Samples are collected at most stations twelve times each year to enable accurate annual flux estimates for most constituents measured. Increased frequency of measurements is required to estimate seasonal or monthly fluxes with greater certainty and also to capture episodic events. This monitoring network does not address delivery of nutrients by smaller rivers to many local coastal ecosystems affected by hypoxia.

Monitoring of inland watersheds is conducted by a diverse group of local programs. However, as indicated above, the design limits their applicability to coastal assessments. USGS also monitors inland watersheds in selected major river systems that drain to the coast, using the same sampling design suitable for estimating annual nutrient fluxes. This monitoring is conducted in the MARB and in the Chesapeake Bay drainages. These data also are used to develop watershed models described in the following section of this report.

Within the MARB, measurements are made at 19 inland stations. The delivery of streamflow and nutrients from the Mississippi River basin to the Gulf of Mexico for the first nine months of the water year (October through June) is reported each year in July to enable Gulf scientists to analyze and forecast the size of the hypoxic zone in the northern Gulf of Mexico (Figure 10b). The hypoxic zone is estimated each year in late summer. In addition, a summary of nutrient loads for all Mississippi River stations for the period of record through 2006 is available on the Internet (http://toxics.usgs.gov/ hypoxia/mississippi/nutrient _flux_yield_est.html). This dataset is updated annually (Figure 15).

3.3.2. Modeling Nutrient Sources, Fate, and Transport

Watershed and nutrient transport, yield, and flux modeling efforts are being undertaken (EPA, NOAA, USDA, USGS) to identify and manage the most significant nutrient sources within watersheds draining to coastal areas experiencing hypoxia. These modeling efforts include the SPAtially Referenced Regressions On Watershed Attributes (SPARROW) Model (http://water.usgs.gov/nawqa/ sparrow/), the Soil and Water Assessment Tool (SWAT) Model (http://www.brc.tamus.edu/swat/), and the Regional Nutrient Management Model (ReNuMa, Hong and Swaney 2007; http://www. eeb.cornell.edu/biogeo/nanc/usda/renuma.htm).

The SPARROW model (Box 8) is a hybrid mechanistic/empirical, basin-scale simulation model developed by the USGS (Smith et al. 1997).

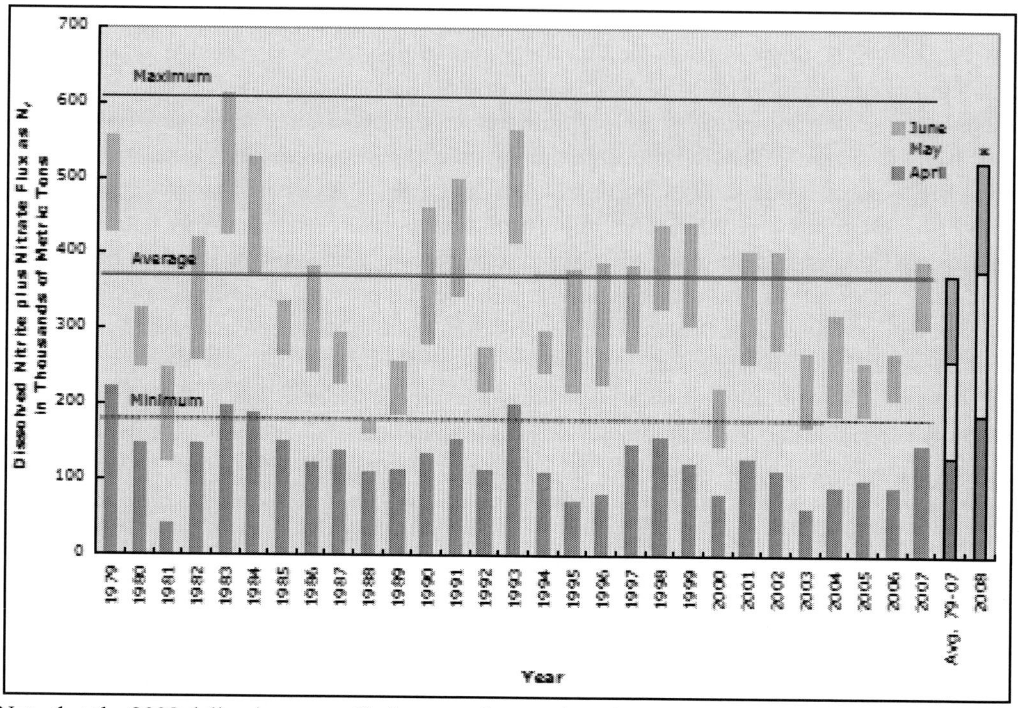

*Note that the 2008 deliveries are preliminary as they are based on provisional data.
Source: Brent Aulenbach, USGS, http://toxics.usgs.gov/hypoxia/mississippi/oct_jun/ graphics.html.

Figure 15. Estimated nitrate delivery to the Gulf of Mexico for April, May, and June 1979 - 2008. Maximum, minimum, and average fluxes are determined through 2007.

Box 8. Application of SPARROW for Reducing Nutrients to the Gulf of Mexico

An improved SPARROW model published in 2008 (Alexander et al. 2008, USGS) found that much of the nitrogen and phosphorus delivered to the Gulf originates from certain regions and watersheds of the Mississippi River Basin, including many watersheds in the Central Mississippi and Ohio, which contribute more than 50% of the nutrients despite accounting for less than 30% of the total drainage area (see inset). The study reveals new details about sources of phosphorus, indicating that animal manure on pasture and range lands contributes nearly as much phosphorus as cultivated crops, 37% versus 43%, respectively. The study reports that 66% of nitrogen originates primarily from cultivated crops, mostly corn and soybean, with animal grazing and manure contributing only about 5%. Atmospheric contributions also are important, accounting for 16% of nitrogen. These findings influenced the 2008 decision by the Mississippi River/Gulf of Mexico Watershed Nutrient Task Force to call for reductions in phosphorus, in addition to nitrogen, in an effort to control Gulf hypoxia.

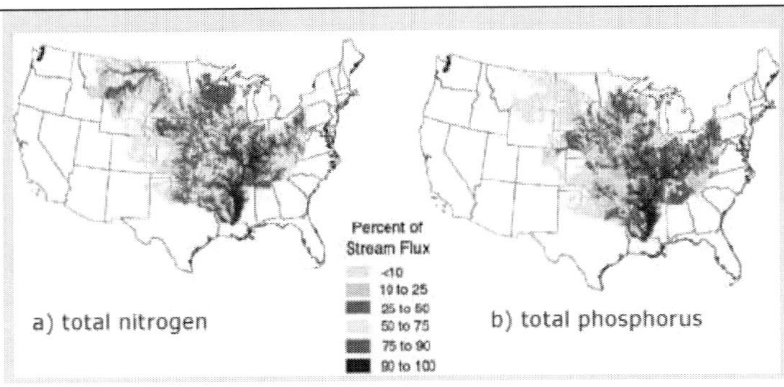

Percentage of stream nutrient load delivered to the Gulf of Mexico: (a): total nitrogen; (b) total phosphorus. (Alexander et al. 2008).

It was developed to cover the entire conterminous United States. Since then, applications have been developed for watersheds of a wide range of scales, with data input refinements appropriate for the specific scale of application. It has been used to estimate nutrient loads and concentrations in streams and to determine the proportion of stream loads that are derived from major sources of nutrients, including land use, chemical use, and human activities (e.g., agriculture, atmospheric deposition, human wastes). The model accounts for the effects of climate, topography, soils, and the effects of aquatic ecosystems to determine nutrient transport in watersheds.

Formal estimates of uncertainties in the stream nutrient loads and source contributions are also reported. The model has been applied to small and very large watersheds in the United States and internationally to assess nutrient sources and loads to the Mississippi River basin (Alexander et al. 2000, 2008), New England watersheds (Moore et al. 2004), the Chesapeake Bay watershed (Preston and Brakebill 1999), New Zealand river basins (Alexander et al. 2002; Elliott et al. 2005), and North Carolina coastal watersheds (McMahon et al. 2003). Many of these applications have demonstrated the particular utility of the model for quantifying the long-distance transport and delivery of nutrients to sensitive downstream locations such as estuaries.

Federal and state environmental managers have used the SPARROW model to assess nutrient sources in streams, including its use for targeting nutrient reduction strategies in the Chesapeake Bay watershed (Preston and Brakebill, 1999) and in waters of the State of Kansas (Kansas Department of Health and Environment, 2004), and for developing total maximum daily loads (TMDLs) in the Connecticut River basin (NEIWPCC, 2004). Regional nutrient models are currently under development in selected regions with the goal of refining the national SPARROW model as part of the USGS National Water Quality Assessment (NAWQA) Program.

The SWAT model was developed by the USDA Agricultural Research Service (ARS; Neitsch et al. 2004; Arnold et al. 1998). It is a physically based, deterministic, continuous, watershed simulation model. It predicts water, sediment, nutrient, and pesticide yields using spatial data on topography, soils, land cover, land management, and weather. The SWAT model has been applied in the Mississippi River basin, including for large watershed-scale applications (Arnold et al. 1999; Anand et al. 2007), to evaluate agricultural nutrient reduction strategies (Santhi et al. 2001; Vache et al. 2002; Hu et al. 2007), address tile-drained cropland (Du et al. 2006; Green et al. 2006), assess the impacts of climate change on nutrient export (Jha et al. 2006), and predict stream flow processes of a forested watershed in coastal South Carolina (Amatya et al. 2008).

A study comparing SWAT and a statistical approach based on SPARROW in the United Kingdom (Grizzetti et al. 2005) suggested using SPARROW as a screening tool for identifying sources, and SWAT for testing management practice scenarios, but found that both models were useful for estimating nitrogen loading in aquatic environments. Similarly, ARS and the NRCS of USDA have developed a system of computer models, the Agricultural Non-Point Source Pollution Model, to predict nonpoint source pollutant loadings in agricultural watersheds.

This model contains a continuous simulation surface runoff model designed to assist in developing BMPs and TMDLs, as well as risk and cost/benefit analyses.

Box 9. EPA Works Closely With States to Develop and Adopt Nutrient Criteria

- Direct assistance to states close to adopting criteria, particularly permitting implementation questions. In this regard, EPA's Office of Science and Technology (OST) in the Office of Water works to develop tools to help states calculate the economic and ecological benefits of adopting water quality standards. OST has provided scientific peer review of proposed water quality standards packages for Vermont and draft nutrient criteria for Colorado. A cooperative assistance grant from EPA's Gulf of Mexico Program to the Mississippi Department of Environmental Quality is assisting MS in all aspects of nutrient criteria development. MS will then lead the other Gulf of Mexico states in nutrient criteria setting for their own states through the Gulf of Mexico Alliance initiative.
- Building capacity of States gathering and analyzing data. Over the past three years, OS1 has provided $2.5 million to support state nutrient criteria work. In 2008, EPA is analyzing nine nutrient data sets that will be used for nutrient criteria. EPA OS1 also developed a number of webcasts that provided expert technical guidance on nutrient criteria issues, with examples from states that have successfully developed nutrient criteria.
- Building a science-based foundation for developing new Clean Water Act water quality criteria for estuaries, wetlands and Large Rivers: OS1 published a wetlands criteria technical guidance manual, developed case studies for two estuarine criteria efforts (Yaquina Bay, Oregon and Pensacola Bay, Florida) in collaboration with EPA's Office of Research and Development, and are supporting four states (Maine, New Hampshire, Florida, and California) who are making progress on estuarine criteria. OS1 is also supporting nutrient criteria development efforts on several large rivers: the Big Sioux (South Dakota), the Missouri River, the Red River, and the Ohio River.

The ReNuMA model is a watershed model that enables the examination of various scenarios for reducing nitrogen losses from the landscape (EPA). Components of the model incorporate biogeochemical complexities and the impact of varying management practices on reducing nutrient loads.

Continual development of the model has included an improved ability to incorporate atmospheric deposition and improve characterization of on-field nutrient management techniques to allow for more accurate simulation of varying management scenarios (NOAA).

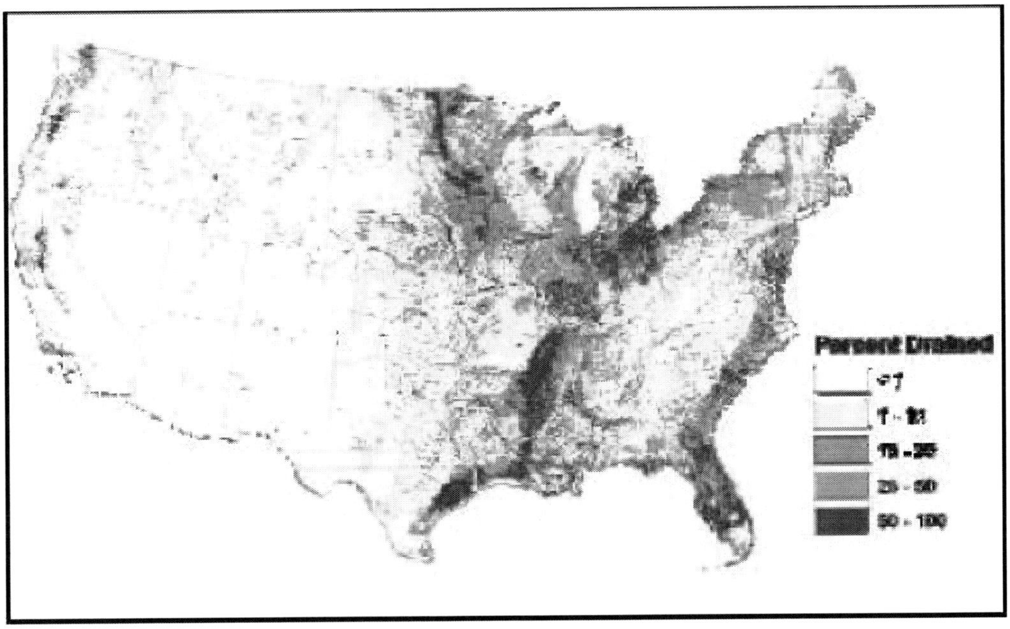

Figure 16. Percent of the United States drained by artificial means such as tile drains (Jaynes and James 2007).

3.4. Assessing, Managing, and Reducing Nutrient Inputs to Coastal Ecosystems

This section discusses tools or management measures that are being studied and/or used to control or reduce nutrient delivery to coastal ecosystems. Nutrient fluxes to coastal ecosystems can primarily be reduced in two ways: (1) reducing nutrient inputs to rivers and streams, including losses of nutrients from farmlands and loads from point sources, urban and suburban nonpoint sources, and atmospheric deposition; and (2) restoring and enhancing natural denitrification and nutrient retention processes via restoration of riverine and coastal wetlands and in some areas, diversions of stream waters to coastal wetlands.

3.4.1. Reducing Nutrient Inputs to Rivers and Streams

3.4.1.1. Establishing Nutrient Criteria

The Clean Water Act directs states and tribes to designate uses for their waters and adopt water quality criteria to protect those uses. Numeric nutrient standards are an important component to protecting water quality from excessive algal blooms and hypoxia because they drive water quality assessments and watershed management, facilitate development of TMDLs, provide quantitative targets to support nutrient trading programs, facilitate writing protective National Pollutant Discharge Elimination System permits, provide increased effectiveness in evaluating success of nutrient runoff minimization programs, and provide objective water quality baselines against which to measure environmental progress. Nutrient criteria benefit not only the local waters in which actions are focused, but also the receiving coastal waters. EPA's Office of Science and Technology (OST) in the Office of Water (OW) is providing both technical and financial assistance to states to help them develop and adopt numeric nutrient standards (Box 9).

3.4.1.2. Tools and Management Strategy Development for Reducing Nutrient Loadings

Reducing nutrient loadings to streams and rivers surrounding and downstream from agricultural lands has been an active area of research for USDA. The re-engineering of the Midwest over the past 50 years with tile drainage systems that allow farmers to control subsurface water levels has benefitted US agriculture through increased yields, but has negatively affected water quality by speeding water and its solutes—such as nitrogen, phosphorus, pesticides, and sediment—into streams and rivers without allowing natural elimination processes to occur (Jaynes and James 2007) (Figure 16). As a result, USDA has made significant investments in conservation programs that remove marginal land from production, conservation improvements in working lands that reduce erosion and control nutrients, and the establishment of wetlands. In addition, USDA has worked with farmers and ranchers to encourage the use of buffer strips and improved tillage practices that are designed to prevent water quality impairments.

Since nutrients often leave agricultural fields through drainage systems (e.g., tile drains, ditches), significant research by the USDA has focused on fine-tuning both surface and subsurface drainage applications (Box 10). In the Chesapeake Bay watershed, it has been estimated that nitrogen losses could be reduced by 40% by lowering the water table during planting and harvesting. Other modified surface drainage systems have the potential to reduce phosphorus by 30-50%, and nitrate and ammonium by 22-64%. In the Gulf of Mexico watershed, researchers have discovered that planting cool-season perennial forages such as alfalfa and grass directly over subsurface tile drains removes nitrate from the water table.

Box 10. Soil Drainage Research in Ohio

Scientists at USDA ARS' Soil Drainage Research Unit in Columbus, Ohio, have developed an innovative agricultural water management system called a Wetland Reservoir Subirrigation System (WRSIS), comprised of a wetland and a water storage reservoir linked to a network of subsurface pipes used at different times to either drain or irrigate crops through the root zone. Runoff and subsurface drainage are collected from cropland into a constructed wetland. Natural processes in the wetland treat the water by removing some of the nutrients, pesticides, and sediment. The water is then routed to a storage reservoir and held until needed to subirrigate the crops during dry parts of the growing season. The storage reservoir also provides a further opportunity for sediment and adsorbed nutrients to settle out of the water. Most of the time, a WRSIS operates as a closed loop, thus restricting offsite water release. Benefits of these systems include: (1) enhanced crop yields; (2) reduced offsite release of nutrients, pesticides, and sediment; (3) increased wetland vegetation and wildlife habitat; (4) carbon sequestration in soil; and (5) the potential to decrease flooding downstream (http://www.ars.usda.gov/ Research/docs.htm? docid=14999).

Other options being explored include deep chiseling and the use of wood chips in drainage ditches to promote denitrification.

Additionally, in 2003, NRCS and ARS (USDA) formed the Agricultural Drainage Management Systems (ADMS) Task Force, in concert with university research and extension personnel. The task force has since expanded to include scientists from many local, state, and Federal agencies, including EPA and USGS. The mission of the ADMS Task Force is to reduce the loss of nitrogen and phosphorus from agricultural lands through drainage water management.

The Task Force focuses on an eight-state region—comprised of Minnesota, Wisconsin, Iowa, Missouri, Illinois, Michigan, Indiana, and Ohio—that contains more than 50 million acres of surface- and subsurface-drained cropland. The Task Force works with farmers, contractors, and agricultural advisors to: 1) implement improved agricultural surface and subsurface drainage in both new and retrofitted systems; 2) reduce nitrate loads in drain outflow, a major source of poor stream water quality and hypoxia in the northern Gulf of Mexico; and 3) improve the efficiency of production and economic returns through managed surface and subsurface agricultural drainage. Implementation of managed drainage practices (NRCS Practice 554) in the Midwest began in 2004.

Using demonstration sites, the Task Force educates local sponsors and farmers and works with them to implement managed drainage on their own land (http:// extension.osu.edu/~ usdasdru/ADMS/ADMSindex. htm).

USDA has invested in developing a variety of strategies and technologies for more effective management of fertilizer applications. USDA developed the Nitrate Leaching and Economic Analysis Package model to evaluate the sensitivity of different regions to nitrate loss. USDA ARS developed the Phosphorus Index, an assessment tool that has been adopted by USDA NRCS, EPA, and various state agencies as the basis for their comprehensive phosphorus management plans. Adoption of this technology has been estimated to reduce annual phosphorus loadings in water by 56 million pounds and sediment by 2.1 billion pounds. Economic benefits to society have been estimated at greater than $600 million dollars per year.

Considerable research has also focused on developing sensors to detect the water and nitrogen status of crops, which allows growers to target fertilizer and water applications to specific crop needs. By reducing unnecessary fertilizer applications, this approach is also more economical. The sensors, some of which have been developed through the Small Business Innovation Research Program grants to private industry, can be either ground-based or airborne.

Ground-based sensors can be mounted on central pivot or other automatic irrigation systems. Researchers are also exploring how to reduce nitrogen and phosphorus release from fertilizers. The benefits of crop rotation, specifically the planting of soy before corn, are being used to justify applying less chemical fertilizer to corn since planting soybeans builds up a nitrogen credit.

Farmers can save up to $30 per hectare with this approach. Finally, cover crops are a proven technology for the eastern and southern states, but are still being researched and introduced to Midwest farmers.

Cover crops (Box 11) may be the best technology for simultaneously reducing nitrogen, phosphorus, and sediment from row crops. Further, USDA's Conservation Reserve Program (CRP) is a voluntary program available to agricultural producers that protects environmentally sensitive land through planting of resource-conserving, long-term vegetative covers (Box 12).

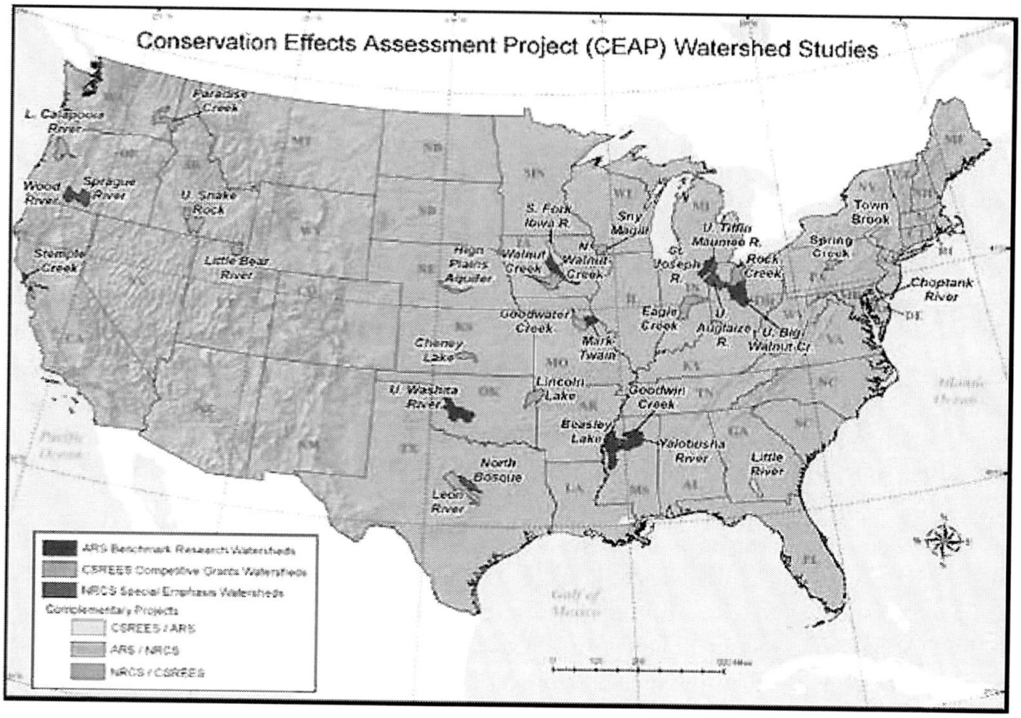

Figure 17. 2007 Map of CEAP projects, many of which are focused on the Mississippi River Watershed (http://www.nrcs.usda.gov/technical/nri/ceap/).

Box 11. Monitoring Winter Cover Crop Performance from Space

Planting winter cover crops is seen as an important management practice for reducing agricultural nutrient flows into the Chesapeake Bay and improving ecosystem health. The State of Maryland doubled its budget for its cover crops cost share program from $9 million to $18 million in 2008. This was enough for farmers to plant 500,000 acres of cover crops after the fall 2008 harvest. USDA scientists at the Beltsville Agricultural Research Center have formed a strong partnership with the Maryland Department of Agriculture (MDA) for the evaluation of winter cover crop performance on the Eastern Shore of Maryland which is closely linked to the Bay. The evaluation uses a unique combination of remotely sensed data and program implementation information such as field location, planting date, cover crop type, and planting method, all supplied by participant farmers during program enrollment. Based on this information, detailed evaluation of cover crop program performance for growth and nutrient uptake can be obtained for all enrolled cropland fields within a region based on analysis of satellite images.

This powerful new approach has led to improvements in the operation of the MDA program for cover crops. It has also demonstrated how satellite data can be used routinely to implement and monitor this important state conservation program. There are plans to jointly develop an operational cover crop implementation and monitoring tool based on technologies developed by this collaboration. This research effort is part of the Choptank River Watershed CEAP project.

Photo: Ben Longstaff, www.eco-check.org.

Box 12. USDA Conservation Research Program Benefits in the Mississippi River Basin

In the Mississippi River Basin, the CRP presence is estimated to confer the following benefits annually:

- Reduced nitrogen runoff: 295 million pounds
- Reduced phosphorus runoff: 66 million pounds
- Reduced sedimentation: 100 million tons
- Protected/restored wetlands: 1.2 million acres
- Carbon dioxide sequestered: 12 million metric tons
- Enhanced wildlife habitat

Based on a rich dataset of observations and an improved, process-based simulation model (called the Root Zone Water Quality Model or RZWQM), ARS scientists have summarized and quantified the effects of management on nitrogen losses to surface waters through a series of papers published as a special issue of the scientific journal *Geoderma* (Ahuja and Hatfield 2007). Both water and nitrogen losses from tile-drained agricultural systems in the Midwest are addressed, as influenced by a variety of management practices. This special issue is a key resource for future efforts to control the magnitude and extent of the hypoxic zone in the Gulf of Mexico by reducing nitrogen loadings from tile-drained agriculture in the Mississippi River basin. Also, USDA's Forest Service is developing practices to improve the capacity of nutrient-enriched watersheds to reduce nutrient loads in surface water and groundwater before they enter coastal waterbodies. For example, the creation of "hot spots," places in a watershed where denitrification could be enhanced, was shown to be an effective way to reduce the amount of nitrogen leaving urban watersheds (Strosneider et al. 2007, Pouyat et al. 2007), resulting in reduced nutrient loading of coastal waters.

In late 2002, the Office of Management and Budget asked NRCS (USDA) to quantify the benefits of Farm Bill-funded conservation programs. In 2003, the Conservation Effects

Assessment Project (CEAP) (Figure 17, Box 11) was initiated by NRCS along with other USDA agencies (ARS; Cooperative State Research, Education, and Extension Service or CSREES; and FSA) to improve scientific understanding of the effects of conservation practices at the watershed scale, and to estimate conservation effects for reporting at regional and national scales. The results are being used to manage agricultural landscapes for environmental quality. CEAP has grown into a USDA-led multi-agency, multi-resource effort (Maresch et al. 2008). Duriancik et al. (2008) provide a detailed description of the approach of CEAP, and of accomplishments during its first five years. Specific details of ARS research accomplishments during the first five years of CEAP are collected in a special issue of the *Journal of Soil and Water Conservation* (2008). ARS completed development of the Sustaining the Earth's Watersheds Through Research, Data Analysis, and Synthesis (STEWARDS) database— an agency-supported repository for CEAP datasets collected as part of ARS' Benchmark Watershed Research Network (available at http:// arsagsoftware.ars.usda.gov/stewards/).

3.4.2. Restoring and Enhancing Natural Nutrient Retention Processes to Reduce Pollution of Aquatic Environments

Nutrient fluxes to coastal ecosystems can be reduced when nutrients already in stream systems are removed by natural processes that occur in riverine wetlands, certain types of reaches of river systems, and coastal/distributary wetlands. Existing and proposed freshwater diversions of the Mississippi River into the estuaries of southeastern Louisiana to mitigate land loss (from many factors including subsidence, salinity intrusion, etc.) have the ancillary benefit of removing nutrients.

The USGS SPARROW model was used to demonstrate that stream channel geometry can affect nitrogen removal in streams in the Mississippi River basin (Alexander et al. 2000). Small streams were found to have a significantly higher nitrogen removal rate, while larger streams acted more like conveyors, transporting the nitrogen load downstream. Therefore, the proximity of nitrogen inputs to larger streams is an important consideration for management actions because it influences whether the nitrogen in a stream will be carried to the Gulf of Mexico or removed.

The Riparian Ecosystem Management Model (REMM) simulates hydrology, nutrient dynamics, and plant growth in land areas between the edge of fields and a waterbody. REMM output allows designers to develop buffer systems to help control nonpoint source pollution (USDA). Analysis of the optimal composition of riparian buffers for nutrient removal has led to various design criteria, including the recommendation that trees or forests be part of the buffer, and analysis of the effectiveness of various grass species as alternatives. The grass analysis determined that all grasses except timothy were beneficial for nutrient removal (USDA).

Some of the natural processes that reduce nutrient fluxes to receiving waters include: 1) restoration of riverine wetlands by the US Fish and Wildlife Service (USFWS) and USACE, 2) management of river flows to increase conditions conducive to increased nutrient reductions, such as lock and dam management on the upper Mississippi River, and 3) coastal diversions onto the Mississippi River delta at projects like the Caernarvon Freshwater Diversion Project. Similarly, flow management on the Mississippi River that diverts 30% of flow down the Atchafalaya River provides increased opportunity for nutrient removal in extensive distributary wetlands in the lower reaches of that river. USDA created more than

12.1 million acres of new wetlands in 2005 and 2006 as a result of research showing that water leaving a wetland had only a fraction of the nutrients (18%) than in the subsurface water entering the wetlands from agricultural fields. The wetlands program within USDA CRP was augmented by the 2008 Food, Conservation, and Energy Act, with a specific emphasis on encouraging producers to construct wetlands to remove nitrogen before it enters streams.

The strategic integration of perennials into agro-ecosystems provides greater benefits than their spatial extent suggest (Schulte et al. 2006; USDA Forest Service). Current research at the field, watershed, and landscape scales in the upper Mississippi River basin is focused on several components that contribute to the mitigation of nutrient loading to the Mississippi River. Researchers have employed simulation models to extrapolate results from small-scale studies over broad scales. Also, research is being conducted on the funding and social mechanisms needed to work with private landowners because, to be successful, the research must be participatory. The USDA Forest Service's Center for Bottomland Hardwood Research in Mississippi is a major partner in the effort to restore agricultural lands to bottomland hardwood ecosystems in the Mississippi Valley. An important part of the program is developing and delivering technology on how to design and install forested buffers to protect water quality.

The USDA Forest Service has conducted extensive research on reforestation restoration techniques, their costs, and expected benefits (Gardiner and Oliver 2005). The majority of reforestation projects in the Lower Mississippi Alluvial Valley (LMAV) use low intensity (less expensive, but less assured in obtaining desired benefits) methods that were established through the CRP and the Wetland Reserve Program (Gardiner and Oliver 2005). These programs provide subsidies for landowners willing to convert agricultural land to forest, range from ten year agreements to perpetual easements, and reimburse landowners for 75% to 100% of the costs of approved practices. By 2007, reforestation reclaimed an estimated 200,000 ha in the LMAV (Gardiner and Lockhart 2007) and 75% of this land is privately owned (Gardiner and Oliver 2005).

A tool that will help achieve that goal is a decision support system that can identify the geographic areas with the greatest potential for restoration and provision of related ecosystem services. The Eco-Assessor, a decision support system developed by the USGS and EPA (Davis et al. 2002), can help planners and managers target areas with the greatest potential for nutrient removal from river water by increasing the weight of the most important functions in the model (water quality) and reducing the weight of less important functions (habitat, hydrology, and restorability). This can allow managers to target areas where the probability of restoration success is highest, and eliminate from consideration areas with less potential. This can sustain public support for wetland restoration by having the most accurate information available to the public.

3.5. Economic Assessments

Determining the economic impacts of coastal hypoxia and the costs and benefits of nutrient reduction within watersheds has proven difficult and complex. However, some progress has been made and efforts are ongoing. Within the watershed, the USDA Economic Research Service (ERS) evaluated the economic costs of reducing nutrient loads to the Gulf of Mexico for the original Committee on Environment and Natural Resources (CENR) study (Doering et al. 1999). Since then, ERS has provided some additional insights on policy options for reducing nutrient loads to the Gulf. In 2001, ERS compared source reduction (fertilizer management) and interception strategies (wetland restoration) for controlling nitrogen loss in the Mississippi River basin (Ribaudo et al. 2001). ERS found that a standard on fertilizer applications was more cost-effective than restoring wetlands up to a particular level of total nitrogen loss reduction (about 26% reduction). Beyond this point, wetland restorations are more cost-effective.

In 2005 ERS assessed the economic benefits of a water quality trading program between point and agricultural sources in the Mississippi River basin (Ribaudo et al. 2005). Allowing trades between point and agricultural sources was found to reduce overall nitrogen abatement costs. A geographically extensive trading program was also found to raise crop prices, increase production, erosion, and nutrient loss outside of the basin.

Additionally, there are ongoing efforts to link the watershed models with the USDA REAP (Regional Environment and Agricultural Programming) economic model developed by USDA (NOAA). This linkage will allow for analyses of the economic costs of varying management scenarios and the corresponding benefits in reducing nutrient inputs

As indicated in Section 3.3.2, limited knowledge of the ecological impacts of hypoxia on living resources has restricted analyses of the economic impacts of hypoxia to coastal communities. However, a recent bioeconomic model developed to examine hypoxia in the Neuse River estuary in North Carolina indicates the economic impacts of hypoxia may be high (NOAA). The model indicates that reductions in the spatial and temporal extent of hypoxia could increase profits to shrimpers, assuming no additional entry into the fishery was permitted. Preliminary estimates from this model have suggested that a 30% reduction in the number of hypoxic days in the Neuse River estuary could potentially increase shrimper profits by $2.5 million annually. As knowledge of the population and physiological impacts of hypoxia grows, it is anticipated that the economic consequences will become clearer.

4. Future Research Directions and Interagency Coordination

4.1. Introduction

Federal research programs are addressing many aspects of eutrophication, nutrient enrichment and hypoxia. Despite decades of research, however, management efforts to reduce nutrients—particularly from nonpoint sources—and their adverse impacts on coastal ecosystems have not made significant headway, in part due to increased development and population in coastal watersheds. The task of managing nutrients and associated hypoxia within and across diverse landscapes and political jurisdictions is complicated by a growing, but still incomplete, understanding of the status of eutrophication of coastal ecosystems and the processes that control water quality responses to excess nutrients. But there are significant opportunities for advancement.

Scientific data and tools are, as of yet, inadequate to fully inform management actions directed at hypoxia in all of the coastal environments impacted by hypoxia. Ideally, such tools would allow managers to readily establish reasonable science-based goals by quantitatively linking management actions to nutrient loads, nutrient loads to coastal water quality, and coastal water quality to ecological outcomes of concern.

This chapter brings together conclusions of a number of reports that independently address research endeavors that could best improve management of hypoxia in US coastal waters. These reports resulted from deliberative, inclusive, and cooperative efforts or workshops focused on determining research priorities. Research priorities related to hypoxia in the northern Gulf of Mexico have received more concerted Federal deliberation recently than priorities for other systems and have been addressed in at least five major reports. These include 1) the 2008 Gulf Hypoxia Action Plan (Mississippi River/Gulf of Mexico Watershed Nutrient Task Force 2008), 2) the Hypoxia in the Northern Gulf of Mexico, An Update by the EPA Science Advisory Board (US EPA 2007), 3) the Gulf of Mexico Hypoxia Monitoring Implementation Plan (Gulf of Mexico Hypoxia Monitoring Implementation Plan Steering Committee 2009), 4) the Science Strategy to Support Management Decisions Related to Hypoxia in the Northern Gulf of Mexico and Excess Nutrients in the Mississippi River Basin (Mississippi River/Gulf of Mexico Watershed Nutrient Task Force 2004), and 5) Nutrient Control Actions for Improving Water Quality in the Mississippi River Basin and Northern Gulf of Mexico (NRC 2008). The Lake Erie Research Planning Workshop Report (NOAA 2004) addressed research needs for Lake Erie, including priorities for addressing hypoxia. The introduction to the special issue of the Journal of Experimental Marine Biology and Ecology on the Ecological Impacts of Hypoxia on Living Resources (Kidwell et al. 2009) provides a framework for guiding research aimed at assessing the effects of hypoxia on living resources. Broad guidance regarding research priorities has been suggested by national assessments such as the National Estuarine Eutrophication Assessment (Bricker et al. 2007). Although research is clearly being done to support management of eutrophication and hypoxia in systems, such as Chesapeake Bay, Long Island Sound, Narragansett Bay, and Hood Canal, recent documentation of an overall research strategy for these coastal ecosystems does not exist.

Some of the identified research needs are common across coastal systems or regions of the country, but there are also differences. These differences result from variability in physical and ecological properties and in the research carried out to date. Four main categories of research needs are identified in this report and discussed below. Within each category, system-specific or regional differences in research needs are examined.

4.2. Improved Characterization and Quantification of Hypoxia

Effective management of hypoxia requires that the causes, extent, and severity of hypoxia be adequately understood and quantified. If the causes are not thoroughly understood, effective management actions cannot be formulated. Adequate monitoring information is required to know the scope of the hypoxia problem and to understand if hypoxia is expanding or being reduced as a result of management actions. Adequate quantification of hypoxia is needed for individual coastal ecosystems that experience recurrent and severe hypoxia (e.g., the northern Gulf of Mexico) and for the Nation's coastal waters as a whole.

Causes of hypoxia are relatively well understood in some systems (e.g., Chesapeake Bay, Gulf of Mexico, Long Island Sound), but there are many other US systems that have only relatively recently been identified as experiencing hypoxia. The general phenomena of transient hypoxia due to diel-cycling of dissolved oxygen levels can have dramatic consequences, such as fish kills, because of its frequency and location of occurrence. This phenomenon is probably more widespread and problematic than currently documented. Process studies are needed to elucidate system dynamics and clarify causes. For example, improved understanding of how nutrient and sediment dynamics contribute to hypoxia will lead to better predictive models. Further, it is currently hypothesized that some systems have experienced regime shifts (e.g., Gulf of Mexico, Chesapeake Bay) due to nutrient overloading which has the effect of increasing the amount of hypoxia for a given load of nutrients. Determining factors underlying such changes is important for refining models and management actions. Hypoxia models provide the ability to test hypotheses about the causes (i.e., nutrient levels, freshwater flow, stratification, sediment dynamics, etc.). Further model development and transfer to operational use by coastal managers should be pursued.

Monitoring of hypoxia in US coastal waters is currently imbalanced, with some systems monitored very well (e.g., Chesapeake Bay) and others monitored much less frequently. In some cases, a significant effort has been expended, but substantial gaps in knowledge remain because the affected area is very large and costly to sample adequately (e.g., Gulf of Mexico hypoxic zone). Monitoring strategies have often been inadequate to characterize the incidence of hypoxia because time or space scales for sampling are too broad. Further, in some waters, hypoxia occurs periodically and primarily (or exclusively) at night and cannot be captured with regular boat surveys during the day.

4.2.1. Hypoxia Surveys

Hypoxia is often difficult to quantify since there is usually considerable temporal and spatial variability. Multiple surveys within a season are usually required to assess goals or standard in a scientifically valid manner and are critical for measuring progress and supporting long-term adaptive management strategies. For example, the extent of hypoxia in the Chesapeake Bay mainstem is monitored by Maryland, Virginia, and EPA at least monthly throughout the summer. This sampling design allows for an accurate assessment of the extent of the problem and has helped managers set dissolved oxygen criteria for the Bay and track progress toward meeting those criteria. In the northern Gulf of Mexico, however, one mid-summer shelf-wide survey is conducted by the Louisiana Universities Marine Consortium as part of NOAA's NGOMEX research program. Given the considerable variability of hypoxia during the summer in this region, one cruise does not provide the coverage needed to monitor progress toward the hypoxia reduction goal. Regularly scheduled shelf-wide monitoring of the northern Gulf of Mexico during the warmer months of April through September would provide the rigorous quantification of hypoxia that is required to assess the goal on an annual basis. This need has been documented in response to a summit that was held to address the current deficiencies in monitoring the hypoxic zone in the northern Gulf of Mexico (Gulf of Mexico Hypoxia Monitoring Implementation Plan Steering Committee 2009).

Although the Chesapeake Bay and some other systems (e.g., Long Island Sound) have been relatively well-sampled, there is no national program or strategy to systematically assess the extent, duration, frequency, or intensity of hypoxia in US coastal waters. A survey program or coordinated monitoring network designed with an appreciation of the temporal dynamics and likely spatial distribution of hypoxia would provide a more complete representation of the extent of hypoxia nationwide.

4.2.2. Instrumented Observing Systems

Ship-based surveys cannot reasonably address the extent of hypoxia in all US coastal waters due to cost and, particularly, because of the temporal dynamics of hypoxia in some coastal systems. This challenge can likely be met by new technologies for measuring dissolved oxygen, particularly optical sensors, which can be deployed in the field to provide continuous measurements. Sensors are increasingly stable in their calibrations and resistance to biofouling. Their data can be relayed to investigators in near real-time via wireless data networks and other telemetry technologies. Instrumented observing systems require an initial infrastructure investment, but then may offer cost-savings relative to traditional surveys. Moreover, only continuously deployed instruments offer a practical means of characterizing temporal dynamics of hypoxia over a range of time scales, including diel-cycling hypoxia. Finally, *in situ* observing in the Gulf of Mexico presents special consideration. The frequency of hurricanes makes maintaining instrumented observing systems expensive due to the high cost of insurance and replacement costs when devices are lost in major storms; such expense is not feasible for most academic researchers. Only a Federal partnership with state and regional entities is likely to succeed in establishing and maintaining a comprehensive national monitoring capability for hypoxia.

4.2.3. Other Advanced Technologies

Autonomous underwater vehicles, such as gliders and powered underwater vehicles, can carry sensor payloads and provide valuable information on the spatial distribution of hypoxia and related water quality variables. Like moored instrumented observing systems, they may generate cost savings by automating sampling and reducing the need for costly ship and boat operations. Hypoxia cannot be monitored directly via satellite remote sensing, but greater use of remote sensing could provide important data on factors that cause or control hypoxia, including chlorophyll-*a*, water clarity, suspended solids, and potentially other variables. Integration of adequate real-time data streams into operational nowcast-forecast simulation modeling systems (e.g., NOAA PORTS models in Mobile Bay, Narragansett Bay, and elsewhere) has the potential to provide both spatially and temporally resolved characterizations of hypoxia in coastal waters with reasonable and quantifiable accuracy.

4.2.4. Improved Modeling of Hypoxia

Because hypoxia results from interactions between physical and biological processes in time and space, simulation models—especially coupled physical-biological water quality models—are especially useful. Scientists increasingly rely on models to test assumptions about the causes of hypoxia. In the past several decades, the technical difficulty, cost to implement, and computational requirements for running these models and evaluating their output have limited their use to a few relatively well-funded modeling programs, usually working in the highest profile areas (e.g., Chesapeake Bay). Recently, this situation has begun to change dramatically. Desktop computers today are vastly more powerful than they were a few years ago. Powerful parallel computing clusters can also be assembled at reasonable cost. Although the models remain complex, it is now possible for more investigators to consider implementing and using them. New and efficient open-source modeling codes such as ROMS (Shchepetkin and McWilliams 2005) and the Finite Volume Community Ocean Model (FVCOM) (Chen et al. 2006) offer powerful and advanced capabilities such as data assimilation, while also enabling relatively simple tests of concepts (e.g., Hetland and DiMarco 2008). The data needed to support the models can increasingly be obtained, sometimes in real-time or near real-time via internet data portals. It now seems possible that even reasonably complex models could be developed and used to support water quality management and other public needs in many coastal systems, or potentially the entire US coast.

An important tool for oceanographic research and water quality sampling - the CTD rosette. Photo: EPA Gulf Ecology Division.

Another consequence of the lower cost of implementing simulations models is that it is now possible and even desirable for several independent or "competing" modeling efforts to address similar issues for the same geographic location.

This development is a positive one because scientific debate and consensus-building—which promote scientific advancement—cannot occur effectively if there is only a single modeling program. As an example, scientists working to address hypoxia in the northern Gulf of Mexico clearly indicated their preference for using a variety of different models relying on different approaches and assumptions (Mississippi River/Gulf of Mexico Watershed Nutrient Task Force 2004). The Federal government should continue to support modeling efforts by supporting open-source model code development, programs that maintain large-scale modeling efforts, and programs that provide data products that facilitate development of local-scale water quality models.

4.3. Improved Characterization of Impacts of Hypoxia

Concern with hypoxia stems from the associated loss of human and aquatic life uses in affected coastal systems. To set reasonable objectives for management of hypoxia, and to justify and sustain efforts to achieve those objectives, it is important to quantify the impacts, both in terms of aquatic life use and ecosystem services provided to humans. Further, it is important to quantify losses that have or are presently occurring, those that will result from a further increase in hypoxia, and those ecosystem services that could be restored by reducing hypoxia.

A known continuum of hypoxia effects has been empirically related to dissolved oxygen concentrations (e.g., Diaz and Rosenberg 1995; Kidwell et al. 2009), providing guidance on likely effects of hypoxia, even in the absence of direct studies of effects in a particular coastal system.

However, direct and systematic assessments of ecological effects of hypoxia in coastal ecosystems, which are important for accurately evaluating impacts, are relatively rare. Dramatic and visible fish kills associated with hypoxia can garner significant public attention, but more serious effects are often much less obvious, such as reduced production.

In systems subject to increases in hypoxia over time, the public and even scientists can fail to recognize both what has been lost and what could be regained by restoring water quality.

4.3.1. Impacts Surveys

Priority research needs related to impacts of hypoxia in coastal waters focus on:

1. characterizing current and past biological condition in ecosystems affected by hypoxia
2. characterizing chronic, sub-lethal effects on individuals, 3) quantifying effects at the population and community level, and 4) characterizing interactions between hypoxia and other stressors.

Characterizing biological condition of coastal ecosystems, such as via faunal assessments, is one of the most direct ways to evaluate ecological impacts due to hypoxia. Improved faunal monitoring presents an opportunity to increase our understanding.

In the northern Gulf of Mexico, NOAA SEAMAP conducts trawl surveys to assess fish abundance, but does not examine benthic community condition. In some less-studied systems where hypoxia has been documented (e.g., Pensacola Bay, Florida), very little data are available to characterize biological condition, and no substantial attempt has been made to relate condition to the extent of hypoxia.

These data limitations contrast with more extensive studies in the Chesapeake Bay system, where decades of benthic and fish surveys describe the effects of hypoxia.

4.3.2. Assessing Biological Responses and Modeling

Models are invaluable tools for quantifying the effects of hypoxia, sometimes providing insights into biotic effects of hypoxia that would be impossible to resolve from empirical observations alone.

Dr. Peter Eldridge extracts sediment cores to investigate sediment biogeochemical processes in the Gulf of Mexico hypoxic zone. Photo: EPA Gulf Ecology Division.

Studies that provide a detailed understanding of the effects of low oxygen on biota present an opportunity to better support these models and increase what can be learned from them. For example, motile species can sometimes avoid exposure to hypoxia by moving away. Species may also modify feeding and other behaviors, impacting growth, reproduction, and often predator avoidance.

Further research describing these behaviors would support better models to predict effects of hypoxia on individuals, populations, and communities.

Quantification of shifts in energy flows within entire food webs as a result of eutrophication and hypoxia (e.g., Baird et al. 2004) would provide integrated perspectives on the effect of hypoxia on marine life in coastal systems.

Unfortunately, hypoxia is rarely the only stressor in coastal systems. Therefore, better data and models are needed to examine hypoxia interactions with additional stressors such as fisheries exploitation, habitat loss, and invasive species. Finally, development of bio-economic models could help assess effects of hypoxia on economically important fisheries and other ecosystem services, thereby helping provide an improved understanding of the possible economic benefits from nutrient management.

4.3.3. Assessing Economic Impacts

For accurate economic analyses, data are needed that account for the many sources of variability influencing population dynamics and landings. In addition, more knowledge on hypoxia occurrence and distribution as well as data on effects on various ecosystem components are needed. Similar limitations are encountered when trying to estimate economic losses from HABs (Anderson et al. 2000). Long-term monitoring programs that measure dissolved oxygen at proper spatial and temporal scales coupled with ecosystem models described in the previous section are critical for resolving the ecological impacts that then set the stage for assessing the economic impacts attributable to hypoxia.

4.4. Quantifying Nutrient Flux to Coastal Waters

Freshwater inflow and nutrient inputs to coastal waters are the primary causes of eutrophication and hypoxia in coastal waters. These factors remain important areas of uncertainty for coastal science and management. Poor quantification of these factors makes it much harder to relate hypoxia and the associated water quality effects in coastal waters to the specific anthropogenic causes in their drainage basins. Tracking nutrients delivered to coastal waters back to source areas and human activities on the landscape is essential for designing effective management strategies to reduce nutrient losses within the drainage basins and to reduce nutrient delivery to coastal systems. Additionally, it is important to quantify the environmental processes that influence transport and fate of nutrients within basins as they affect ultimate delivery to coastal ecosystems.

4.4.1. Monitoring Nutrient Sources and Fluxes to Coastal Ecosystems

Monitoring streamflow and nutrient fluxes theoretically is straight-forward. Load estimates can be made using well-established regression-based models (Cohn et al. 1992; Runkel et al. 2004) or related methods (Aulenbach and Hooper 2006) that rely on streamflow monitoring data and regular measures of water chemistry. However, the number of gauged rivers is small, and contributing drainage basins can extend over broad areas quite distant from the coast. As a result, significant monitoring is required to provide sufficient data to make adequate estimates of the delivery of streamflow and nutrients to the coast and the source areas and human activities within the basin that affect them.

Although simple in theory, there remains room to improve the statistical methods of load estimation directed largely at reducing the frequency and costs associated with water quality monitoring. Nutrient load estimates with adequate temporal resolution to develop cause-and-effect linkages to the onset, extent, and duration of hypoxia and to support empirical and simulation models of coastal systems currently are not available in the United States. This is largely a result of insufficient frequency of water quality monitoring and inadequate streamflow gauging. Acceptably accurate estimates of loads are needed at monthly or higher frequencies (e.g., weekly). However, increased temporal resolution comes at the expense of increased limits of uncertainty unless there is a commensurate increase in the frequency of sampling.

Continuous real-time nutrient monitoring stations offer a potential solution. These have been pioneered in a few locations such as the mouths of the Mississippi and Atchafalaya Rivers. However, the technology needs further development to be more practical and reliable and to provide data on a sufficient suite of water quality constituents.

Monitoring riverine delivery of nutrients beyond tidal waters to coastal systems has been complicated by difficulties in stream gauging and representative sampling in tidal reaches.

New techniques employing acoustic Doppler current profilers have made these measurements possible and reduced the uncertainty associated with nutrient removal or additions in tidal river reaches (Box 13).

Measurements of point source loadings of nutrients are also needed. For most coastal waters, these are not sufficiently accessible. The Chesapeake Bay Program's Nutrient Point Source Database provides a good example of what should be implemented much more broadly. A consistent, systematic, and comprehensive nationwide approach for estimating nutrient loads to coastal waters from all upland sources (fluvial, diffuse, point source, atmospheric) would be a very valuable tool to support science and management related to hypoxia in US coastal waters. Estimates for atmospheric deposition, both dry and wet, to coastal systems is a source of great uncertainty in the understanding of nitrogen and carbon cycling in watersheds.

4.4.2. Modeling Source Areas, Source Mechanisms, and Trends in Nutrient Loads

Modeling is a valuable tool for interpreting data across a broad range of scales and employing a range of approaches from purely deterministic simulation modeling to statistically based empirical approaches.

The interpretation of trends in streamflow and nutrient loads provides important feedback to adaptive management. Separation of natural variations from those induced by human activities, such as changes in land use practices and management actions to reduce nutrient loses, can be extremely challenging. New statistical models are needed to implement such time series analyses.

There are insufficient resources to monitor streamflow and nutrient loads at all inland watersheds. As a result, monitoring is done in watersheds representative of others with similar climate, land use, other measures of nutrient inputs, wetlands, and management actions. In turn, the models help identify the representative watersheds. The models then extend the monitoring data across the entire basin to show the different contributions geographically and to identify the relative contributions from different anthropogenic sources. Such spatial information should be considered when management actions are implemented so that resources are invested where they will have the greatest benefit. Although the SPARROW model has been applied at a national scale (Smith et al. 1997), as well as regional (e.g., Preston and Brakebill 1999, Moore et al. 2004) and smaller scales, one limitation of SPARROW is that it has been applied to estimate average annual loading. Other approaches are needed where seasonal or interannual differences must be resolved. Isolated implementation of models, such as Hydrological Simulation Program-Fortran, can meet this need, but it has not been developed for most of the Nation's estuaries.

Ancillary data on human activities across the landscape that affect nutrient inputs are critical to the development of models that extend monitoring data across large contributing drainage basins. These ancillary data include information on chemical use, land use, and information on the types and distribution of management actions intended to mitigate nutrient losses to streams. The data include type of crop or animal agriculture; the extent, pattern, and intensity of agricultural drainage; the extent of riverine ecosystems that affect instream nutrient concentrations; and improved estimates of direct nutrient inputs (atmospheric deposition and point source discharges). Furthermore, to make accurate projections of future trends, research is needed to improve understanding of the consequences of long-term drivers, such as climate change, changes in cropping systems, extent of tile drainage, and biofuels production.

Groundwater also can provide a significant contribution of water and nutrients to coastal waterbodies. It is difficult to quantify these flows and loads. Although ground-water flow system models can improve estimates of these contributions, they are difficult and expensive to develop. Cost-effective models are needed to assess the relative role of ground water in coastal ecosystems and to quantify its contribution where significant.

4.4.3. Nutrient Processes in Watersheds

Quantification of the biogeochemical processes that affect nutrient transport on the landscape and within streams will also support development of sound land and water quality management practices. These processes affect how nutrients are transported from lands with various land use practices (crop agriculture, animal agriculture, urban, suburban) to streams and include various components of the hydrologic cycle (runoff, groundwater seepage, tile drainage, and atmospheric transport). Biogeochemical processes determine nutrient transport and persistence downstream to receiving waters, including nutrient transformation and removal. Ecosystem types that are most effective at reducing instream nutrient loads and the characteristics that make them most efficient must be identified. These include riverine, estuarine, coastal, wetland, and delta ecosystems.

Research is needed on small watersheds to integrate understanding of the source of nutrients; the role of groundwater, surface waters, riparian zones and wetlands in the transport of nutrients to the stream; and the downstream persistence of nutrients. A better understanding is needed of the temporal and seasonal variation in nutrient fluxes and the time lags associated with changes in drivers (increased inputs or mitigation measures) and resulting changes in nutrient loads. The role of various environmental media (sediment and organic matter) in defining nutrient budgets is poorly understood. Improved approaches are needed to characterize water and nutrient balances for land areas to which mitigation measures may be applied to ensure that resources are invested in the highest priority areas and to areas where management actions will have the greatest effect on nutrient loads to receiving coastal waters.

Most research on nutrient cycling in watersheds is driven largely by concerns for science-based management actions that improve local water quality conditions. Typically, this research is not driven by concerns for water quality effects on coastal ecosystems, which can be quite distant— both geographically and culturally. An important goal of research on nutrient processes that is pertinent to hypoxia should be to consider aspects that are important to downstream receiving waters.

Box 13. USGS Deployment of New Instruments to Measure Water Flow and Sediment Flux

USGS is now deploying acoustic Doppler current profilers on a variety of platforms to monitor a range of hydrologic measurements and applications, including computing continuous records of streamflow for tidally or backwater affected streams, measuring velocity fields with high spatial and temporal resolution, and estimating suspended-sediment concentrations.

ADCPs are used on a variety of measurement platforms. (A) Manned boats (photography by Paul [illegible], USGS). (B) Tethered boats (photograph by Kevin Oberg, USGS). (C) Remotely controlled boat.

4.5. Approaches for Reducing Nutrient Inputs to Coastal Waters

This research is directed at developing improved management tools for mitigating excessive nutrient fluxes to coastal ecosystems. It has two major components: 1) *reducing nutrient inputs* – research to identify BMPs for reducing nutrient inputs to streams, including losses from agricultural lands, point source inputs, urban/suburban runoff, and atmospheric deposition; and 2) *enhancing nutrient removal* – research to identify the instream nutrient reductions taking place in wetland areas, including reconstructed wetlands, river reaches amenable to nutrient removal processes, constructed riverine wetlands, and coastal diversions. Process research on nutrient cycling and transport is fundamental to both of these components. Understanding where nutrients are entering the watershed and how much each region is contributing will lead to more cost-effective targeting of nutrient reduction strategies.

4.5.1. Evaluation of Nutrient Reduction Strategies

Evaluations of nutrient reduction strategies should continue in a deliberate and measurable way to promote the most efficient and cost-effective management alternatives. CEAP (see Section 3.4.1.2) is already proving to be a good tool for this purpose and may warrant expansion to more regions of the United States. CEAP and other scientific assessments should be continued to evaluate the effectiveness of current nutrient reduction management techniques, while developing ways to maximize their effectiveness and improve implementation and maintenance, including market-based environmental stewardship programs. Other assessments which should improve nutrient reduction management include: 1) socioeconomic and bioeconomic assessments of alternative nutrient management scenarios through the development of integrated economic and watershed models; 2) more thorough evaluation of the efficacy of dual nutrient (nitrogen and phosphorous) control practices; and 3) evaluation of the most effective ways to market nutrient reduction strategies to farmers. Further, the importance of seasonality and nutrient form should be considered in nutrient reduction strategies.

Finally, as the understanding of climate change impacts becomes more refined, this new knowledge needs to be applied to current nutrient reduction strategies.

4.5.2. Biofuels Research

Research on biofuel life cycles is needed to assess the environmental effects, economics, and resource availability of biofuel feedstocks. Research has already shown fewer environmental impacts if cellulosic feedstocks are used instead of corn. Region-specific biofuel production systems should be encouraged to maximize production while minimizing environmental impacts. "Region-specific" implies a more localized approach to the type of feedstock used for biofuel distillation (e.g., considering what grows best where), thereby reducing distance between farms and biofuel plants and closing the loop on byproducts reuse (e.g., feeding corn mash to live stock).

4.5.3. Local Water Quality Concerns

In inland states, there are inherent difficulties in developing an appreciation for the implications that nutrient inputs to rivers ultimately have on coastal ecosystems. Therefore, it is important to consider both local and regional water quality concerns in nutrient studies and management strategies. Studies should consider the implications of excess nutrients in local streams and estimate the costs in terms of both lost ecosystem services and the local and regional benefits associated with improving local water quality conditions. Excess nutrients in streams, rivers, and lakes can result in a plethora of problems, including unsightly algal blooms, fish kills, nitrate or ammonia toxicity, and contaminated drinking water. The safety standard (maximum contaminant level) for nitrate in drinking water established by the EPA is 10 mg/L (Ward et al. 2005). Nitrate levels at or above this level have been known to cause a potentially fatal blood disorder in infants under six months of age called methemoglobinemia or "blue-baby" syndrome. Therefore, when selecting nutrient management strategies, it is imperative to look for 'win-win' situations that improve both local and coastal water and habitat quality.

4.5.4. Nutrient Management Technology Development

Given the development of new technologies, such as the use of remote sensing for new applications (which include assessment of crop nutrient needs), it is important to continue to develop nutrient management technologies (e.g., cover cropping, siting of BMPs, stormwater controls) to reduce nitrogen and phosphorus inputs to the coasts arising from both agriculture and urban/suburban sources. Achieving higher sequestration of applied nutrients by crops and reducing losses from soils without compromising agricultural yield should be one of the central goals of this research. In general, the science and management practices related to control of nonpoint nutrient sources is far behind that for point sources, such as wastewater treatment plants.

4.5.5. Improve Natural Nutrient Removal Processes

Evaluation and exploration of methods to enhance natural processes that remove nutrients from streams and rivers should be a high priority. Naturally vegetated watersheds, and particularly those that contain wetlands, are highly effective at nutrient retention and are important in both improving and sustaining water quality. Research is needed on how to identify the most promising land areas and methods to restore the ecosystem services provided by vegetated watersheds and wetlands. Also, adaptation research is needed to increase the resiliency of forests, rangelands, and aquatic areas in an effort to mitigate the adverse impacts of climate change. As the understanding of ecosystem processes improves, methods to evaluate ecosystem services and to develop incentives for the public will be needed.

4.6. Existing and Future Coordination

In the immediate future, there is a strong need for improved data access across agencies. Access to data is critical for model development. Coordinated, consolidated, and improved access to data, relevant to particular watersheds and to hypoxia in the coastal zone and for use by watershed and coastal zone managers, should be available across agencies and to the public. NOAA provides data access through the National Oceanographic Data Center (http://www.nodc. noaa.gov/) which requires all NOAA-funded researchers to deposit data within two years of project completion. EPA's STOrage and RETrieval (STORET) data system provides the same function for EPA-funded research (http:// www.epa.gov/storet /about.html). USGS data are stored in the USGS National Water Information System, which has a website for public access to the data (http://waterdata.usgs.gov/nwis). EPA and USGS have made significant progress toward providing scientists and policy makers an easier way to integrate access to their large water quality databases. A common suite of web services allow for the automated sharing of water monitoring data via a common format and terms (http://qwwebservices.usgs.gov/). USDA's ARS has recently developed STEWARDS (http://arsagsoftware.ars.usda. gov/ stewards/) as a repository for data collected as part of ARS' watershed research network activities towards the CEAP croplands initiative. All of the agencies need to improve access and usability and provide an interface among their various databases.

Growing cooperation among Federal agencies and among Federal, state, local, and tribal agencies has enhanced hypoxia monitoring and research capabilities. Most coastal regions with hypoxia problems in the United States now have well coordinated, multi-agency, multi-state management plans with informed, active science and technology advisory committees, many of which evolved from the National Estuary Program initiative. Some of these regional partnerships have existed for decades while others are just forming. For instance, the Chesapeake Bay Program (http:// www.chesapeakebay.net/), which was founded in 1983, engages a range of stakeholders and has been responsible for organizing real, tangible actions to improve water quality in the Bay.

On the other hand, the Puget Sound Partnership, which was just formed in 2007, is a relatively new effort to address the worsening water quality resulting from development of the Puget Sound watershed.

A recently released report (December 2008) from the National Research Council of the National Academies recommends the creation of a Mississippi River Basin Nutrient Control Implementation Initiative (NCII) to be established jointly within the USDA and EPA (http://www. nap.edu/catalog. php?record_id=12544). The NCII would focus on many of the research recommendations outlined in Section 4.6 (e.g., pilot studies on nutrient reduction, analysis of cost-effective nutrient reduction strategies, etc.), but more importantly it would solidify coordination between the two Federal agencies primarily responsible for improving water quality in freshwater US systems.

Ultimately, though, there is no national policy to confront the issues of eutrophication and hypoxia, and no single Federal agency is in charge of coordinating a Federal response. As evidenced in this report, coordinated efforts are required across at least four major agencies (EPA, NOAA, USGS, USDA, as well as others) in order to address both the impacts of hypoxia and its root causes.

Currently, the IWG-4H provides some measure of communication and coordination about hypoxia across agencies.

In addition, interagency communication happens through deliberations in other IWGs which address related issues, such as the sustainable development of biofuels.

ACKNOWLEDGMENTS

Many scientists and managers from Federal and state agencies, universities, and research institutions contributed to the knowledge base upon which this assessment depends. Many thanks to all who contributed to this report, and special thanks to John Wickham and Lynn Dancy of NOAA National Centers for Coastal Ocean Science for their editing work.

REFERENCES

Ahuja LR., Hatfield JL. 2007. Integrating Soil and Crop Research with System Models in the Midwest USA: Purpose and Overview of the Special Issue in Geoderma 140(3): 297-309.

Alexander RB, Smith RA, Schwarz GE. 2000. Effect of stream channel size on the delivery of nitrogen to the Gulf of Mexico: *Nature* 403: 758-761.

Alexander RB, Elliott AH, Shankar U, McBride GB. 2002. Estimating the sources and transport of nutrients in the Waikato River basin, New Zealand: *Water Resources Research* 38: 1268-1290.

Alexander RB, Smith RA, Schwarz GE, Boyer EW, Nolan JV, Brakebill JW. 2008. Differences in phosphorus and nitrogen delivery to the Gulf of Mexico from the Mississippi River Basin, *Environ. Sci. Technol.* 42(3): 822-830.

Aller RC. 1994. Bioturbation and remineralization of sedimentary organic matter: Effects of redox oscillations. *Chemical Geology* 114: 331–345.

Altieri AH, Witman JD. 2006. Local extinction of a foundation species in a hypoxic estuary: integrating individuals to ecosystem. *Ecology* 87: 717-730.

Altieri AH. 2008. Dead zones enhance key fisheries species by providing predation refuge. *Ecology* 89(10): 2808- 2818.

Amatya DM., Haley EB, Levine NS, Callahan TJ, Radecki-Pawlik A, Jha MK. 2008. Calibration and validation of the SWAT Model for a forested watershed in Coastal South Carolina. ASABE Paper no. 083912, St. Joseph, Nich.: ASABE.

Anand S, Mankin KR, McVay KA, Janssen KA, Barnes PL, Pierzynski GM. 2007. Calibration and validation of ADAPT and SWAT for field-scale runoff prediction: *Journal of the American Water Resources Association* 43(4): 899-910.

Anderson DM, Hoagland P, Kaoru Y, White AW. 2000. Estimated Annual Economic Impacts from Harmful Algal Blooms (HABs) in the United States. WHOI Tech Rept. No. 2000-11. Woods Hole Oceanographic Institution, Department of Biology and Marine Policy Center,.Woods Hole, MA.

Anderson RS, Brubacher LL, Calvo LR, Unger MA, Burreson EM. 1998. Effects of tributyltin and hypoxia on the progression of Perkinsus marinus infections and host defense mechanisms in oyster, Crassostrea virginica (Gmelin). *Journal of Fish Diseases* 21: 371-380.

Arnold JG, Srinivasan R. Muttiah RS, Williams JR. 1998. Large area hydrologic modeling and assessment—Part I, Model development: *Journal of the American Water Resources Association* 34(1): 73–89.

Arnold JG, Srinivasan R, Muttiah RS, Allen PM. 1999. Continental scale simulation of the hydrologic balance: *Journal of the American Water Resources Association* 35(5): 1037–1051.

Aulenbach BT, Hooper RP. 2006. The composite method - An improved method for stream-water solute load estimation. *Hydrol. Proc.* 20: 3029-3047.

Baden SP, Loo LO, Pihl L, Rosenberg R. 1990. Effects of eutrophication on benthic communities including fish - Swedish west coast. *Ambio* 19: 113-122.

Baird D, Ulanowicz RE. 1989. The seasonal dynamics of the Chesapeake Bay ecosystem. *Ecological Monographs* 59(4): 329-364.

Baird D, Christian RR, Peterson C, Johnson GA. 2004. Consequences of hypoxia on estuarine ecosystem function: energy diversion from consumers to microbes. *Ecological Applications* 14(3): 805-822.

Boesch DF, Rabalais NN. 1991. Effects of hypoxia on continental shelf benthos: comparisons between the New York Bight and the Northern Gulf of Mexico. In "Modern and ancient continental shelf anoxia" (R. V. Tyson and T. H. Pearson, Eds.), Vol. 58, pp. 27-34. *Geological Society Special Publication*.

Boesch DF, Coles VJ, Kimmel DG, Miller WD. 2007. Coastal dead zones and global climate change: ramifications of climate change for the Chesapeake Bay, pp. 54-70, In: Regional Impacts of Climate Change: Four Case Studies in the United States. Pew Center for Global Climate Change, Arlington, VA.

Breitburg DL. 1992. Episodic hypoxia in Chesapeake Bay: interacting effects of recruitment, behavior, and physical disturbance. *Ecol. Monogr.* 62: 525-46.

Breitburg DL. 1994. Behavioral response of fish larvae to low dissolved oxygen concentrations in a stratified water column. *Marine Biology* 120: 615–625.

Breitburg DL. 2002. Effects of hypoxia, and the balance between hypoxia and enrichment, on coastal fishes and fisheries. *Estuaries* 25: 767-781.

Breitburg DL, Pihl L, Kolesar SE. 2001. Effects of low dissolved oxygen on the behavior, ecology and harvest of fi shes: A comparison of the Chesapeake and Baltic systems, p. 241–267. In N. N. Rabalais and R. E. Turner (eds.), Coastal Hypoxia: Consequences for Living Resources and Ecosystems. Coastal and Estuarine Studies 58. *American Geophysical Union*, Washington, DC.

Breitburg DL, Hondorp DW, Davias LW, Diaz RJ. 2008. Hypoxia, nitrogen and fisheries: Integrating effects across local and global landscapes. *Annual Reviews in Marine Science* 20(4): 329-349.

Breitburg DL, Craig JK, Fulford RS, Rose KA, Boynton WR, Brady D, Ciotti BJ, Diaz RJ, Friedland KD, Hagy JD III, Hart DR, Hines AH, Houde ED, Kolesar SE, Nixon SW, Rice JA, Secor DH, Targett TE. 2009. Nutrient enrichment and fisheries exploitation: interactive effects on estuarine living resources and their management. *Hydrobiologia*: 629: 31–47.

Bricker S, Clement C, Frew S, Harmon M, Pirhalla D. 1996. NOAA's Estuarine Eutrophication Survey. Volume 1: South Atlantic Region. Office of Ocean Resources Conservation Assessment. *Silver Spring*, MD. 50 pp.

Bricker S, Clement C, Frew S, Harmon M, Harris M, Pirhalla D. 1997a. *NOAA's Estuarine Eutrophication Survey.*

Volume 2: Mid-Atlantic Region. Office of Ocean Resources Conservation Assessment. *Silver Spring*, MD. 51 pp.

Bricker S, Clement C, Frew S, Harmon M, Harris M, Pirhalla D. 1997b. NOAA's Estuarine Eutrophication Survey. Volume 3: North Atlantic Region. Office of Ocean Resources Conservation Assessment. *Silver Spring*, MD. 46 pp.

Bricker S, Clement C, Frew S, Harris M, Pirhalla D. 1998a. NOAA's Estuarine Eutrophication Survey. Volume 5: Pacific Coast Region. Office of Ocean Resources Conservation Assessment. *Silver Spring*, MD. 75 pp.

Bricker S, Clement C, Frew S, Harmon M, Harris M, Pirhalla D. 1998b. NOAA's Estuarine Eutrophication Survey. Volume 4: Gulf of Mexico Region. Office of Ocean Resources Conservation Assessment. *Silver Spring*, MD. 78 pp.

Bricker S, Clement C, Pirhalla D, Orlando S, Farrow D. 1999. National Estuarine Eutrophication Assessment. Effects of Nutrient Enrichment in the Nation's Estuaries. NOAA, National Ocean Service, Special Projects Office and National Centers for Coastal Ocean Science, Silver Spring, MD. http:// specialprojects.nos.noaa.gov/projects/cads/ nees/Eutro_Report.pdf

Bricker SB, Lipton D, Mason A, Dionne M, Keeley D, Krahforst C, Latimer J, Pennock J. 2006. Improving methods and indicators for evaluating coastal water eutrophication: A pilot study in the Gulf of Maine. NOAA technical report 20. http://ccma.nos. noaa.gov/news/feature/GulfofMaine.html

Bricker S, Longstaff B, Dennison W, Jones A, Boicourt K, Wicks C, Woerner J. 2007. Effects of Nutrient Enrichment in the Nation's Estuaries: A Decade of Change, National Estuarine Eutrophication Assessment Update. NOAA Coastal Ocean Program Decision Analysis Series No. 26. National Centers for Coastal Ocean Science, Silver Spring, MD. 322 pp. http://ccma.nos.noaa. gov/ news/feature/Eutroupdate.html

Brosnan TM, O'Shea ML. 1996. Long-term improvements in water quality due to sewage abatement in the lower Hudson River. Estuaries 19: 890-900.

Brown CA, Nelson WG, Boese BL, DeWitt TH, Eldridge PM, Kaldy JE, Lee H II, Power JH, Young DR. 2007. An Approach to Developing Nutrient Criteria for Pacific Northwest Estuaries: *A Case Study of Yaquina Estuary*, Oregon. US

Burgents, J.E., K.G. Burnett, L.E. Burnett. 2005. Effects of hypoxia and hypercapnic hypoxia on the localization and the elimination of Vibrio campbellii in Litopenaeus vannamei, the Pacific white shrimp. Biol. Bull. 208: 159-168.

Caddy J. 1993. Toward a comparative evaluation of human impacts on fishery ecosystems of enclosed and semi-enclosed seas. Rev. Fishery Sci. 1: 57-96.

CBC (Chesapeake Bay Commission). 2007. Biofuels And the Bay. Getting It Right To Benefit Farms, Forests And The Chesapeake. http://www. chesbay.state.va.us/ Publications /BiofuelsAndTheBay1.pdf.

CENR (Committee on Environment and Natural Resources). 2000. Integrated Assessment of Hypoxia in the Northern Gulf of Mexico. National Science and Technology Council Committee on Environment and Natural Resources, Washington, DC.

CENR (Committee on Environment and Natural Resources). 2003. An Assessment of Coastal Hypoxia and Eutrophication in US Waters. National Science and Technology Council Committee on Environment and Natural Resources, Washington, DC.

Cerco CF, Noel MR. 2005. Incremental improvements in Chesapeake Bay Environmental Model Package. *J. Environmental Engineering* 131: 745-754.

Cerco CF, Noel MR. 2004. The 2002 Chesapeake Bay Eutrophication Model. US Environmental Protection Agency, Region 3 Chesapeake Bay Program Office, *US Army Corps of Engineers.* EPA 903-R-04-004.

Chan F, Barth JA, Lubchenco J, Kirincich A, Weeks H, PetersonWT, Menge BA. 2008. Emergence of anoxia in the California current large marine ecosystem. *Science* 319: 920.

Chen C, Beardsley RC, Cowles G. 2006. An unstructured-grid finite-volume coastal ocean model (FVCOM) System. Oceanography 19(1): 78-89.

Chesapeake Bay Commission and Commonwealth of Pennsylvania. 2008. Next-Generation Biofuels: Taking the Policy Lead for the Nation. http://www.chesbay.state.va.us/Publications/nexgen%20biofuels1.pdf

Christensen JH, B. Hewitson, A. Busuioc, A. Chen, X. Gao, I. Held, R. Jones, R.K. Kolli, W.T. Kwon, R. Laprise, V. Magaña Rueda, L. Mearns, C.G. Menéndez, J. Räisänen, A. Rinke, A. Sarr, P. Whetton. 2007. Regional Climate Projections. In Climate Change 2007: The Physical Science Basis. Contribution of Working Group I to the Fourth Assessment Report of the Intergovernmental Panel on Climate Change. S. Solomon, D. Qin, M. Manning, Z. Chen, M. Marquis, K. B. Averyt, M.Tignor, and H. L. Miller (eds.), Cambridge University Press, Cambridge, United Kingdom and New York, NY.

Cohn TA, Caulder DL, Gilroy EJ, Zynjuk LD, Summers RM. 1992. The validity of a simple statistical model for estimating fluvial constituent loads—An empirical study involving nutrient loads entering Chesapeake Bay: *Water Resources Research* 28(9): 2353–2363.

Cooper SR, Brush GS. 1991. Long-term history of Chesapeake Bay anoxia. *Science* 254: 992-996.

Cowan JLW, Boynton WR. 1996. Sediment-water oxygen and nutrient exchanges along the longitudinal axis of Chesapeake Bay: Seasonal patterns, controlling factors and ecological significance. *Estuaries* 19: 562-580.

Craig JK, Crowder LB. 2005. Hypoxia-induced habitat shifts and energetic consequences in Atlantic croaker and brown shrimp on the Gulf of Mexico shelf. *Mar. Ecol. Prog. Ser.* 294: 79-94.

Craig JK, Crowder LB, Henwood TA. 2005. Spatial distribution of brown shrimp (Farfantepenaeus aztecus) on the northwestern Gulf of Mexico shelf: effects of abundance and hypoxia. *Canadian Journal of Aquatic Science* 62: 1295-1308.

D'Avanzo C, Kremer JN. 1994. Diel oxygen dynamics and anoxic events in an eutrophic estuary of Waquoit Bay, Massachusetts. *Estuaries* 17(1B): 131-139.

Davis AA, Kleiss B, O'Hara CG, Derby JS. 2002. The development of a decision support system for prioritizing forested wetland restoration areas in the lower Yazoo River Basin, Mississippi. Pg. 47-56 In. Marjorie M. Holland, Melvin L. Warren, John A. Stanturf (eds.). Proceedings of a conference on sustainability of wetlands and water resources: how well can riverine wetlands continue to support society into the 21st century? Gen. Tech. Rep. SRS-50. US Department of Agriculture, Forest Service, Southern Research Station, Asheville, NC. 191 pp.

Diaz RJ, Breitburg DL. 2009. The hypoxic environment. In: Richards JG, Farrell AP, Brauner CJ (eds.), Fish Physiology, Hypoxia. Vol. 27. Academic Press, Burlington, VT. pp.1-23.

Diaz RJ, Schaffner LC. 1990. The functional role of estuarine benthos. pp. 25-56. In: M. Haire and E.C. Krome (eds.). Perspectives on the Chesapeake Bay, 1990. Advances in estuarine sciences. Chesapeake Research Consortium, Gloucester Pt., VA. Rpt. No. CBP/TRS41/90.

Diaz RJ, Rosenberg R. 1995. Marine benthic hypoxia: A review of its ecological effects and the behavioural responses of benthic macrofauna. Oceanogr. *Mar. Biol. Annu. Rev.* 33: 245-03.

Diaz RJ, Rosenberg R. 2008. Spreading Dead Zones and Consequences for Marine Ecosystems. *Science* 321: 926-929.

Doering OC, Diaz–Hermelo F, Howard C, Heimlich R, Hitzhusen F, Kazmierczak R, Lee J, Libby L, Milon W, Prato T, Ribaudo M. 1999. Evaluation of the Economic Costs and Benefits of Methods for Reducing Nutrient Loads to the Gulf of Mexico: Topic 6 Report for the Integrated Assessment on Hypoxia in the Gulf of Mexico. NOAA Coastal Ocean Program Decision Analysis Series No. 20. NOAA Coastal Ocean Program, *Silver Spring,* MD. 115 pp.

Donner SD, Kucharik CJ. 2008. Corn-based ethanol production compromises goal of reducing nitrogen export by the Mississippi River. *Proceedings of the National Academies of Science.*105(11): 4513-4518.

Downing JA, Rabalais NN, Diaz RJ, Zimmerman RJ, Baker JL, Prato R. 1999. Gulf of Mexico hypoxia: Land-sea interactions. *Council for Agricultural Science and Technology*, Report No. 134, 44 pp.

Du B, Saleh A, Jaynes DB, Arnold JG. 2006. Evaluation of SWAT in simulating nitrate nitrogen and atrazine fates in a watershed with tiles and potholes: *Transactions of the American Society of Agricultural and Biological Engineers* 48(4): 949–959.

Duriancik LF, Bucks D, Dobrowolski JP, Drewes T, Eckles SD, Jolley L, Kellog RL, Lund D, Makuch JR, O'Neill MP,

Rewa CA, Walbridge MR, Parry R, Weltz MA. 2008. The first five years of the Conservation Effects Assessment Project. *Journal of Soil and Water Conservation* 63(6): 185A – 197A.

Eby LA, Crowder LB, McClellan CM, Peterson CH, Powers MJ. 2005. Habitat degradation from intermittent hypoxia: impacts on demersal fishes. *Mar. Ecol. Prog. Ser.* 291: 249-261.

Elliott AH, Alexander RB, Schwarz GE, Shankar U, Sukias JPS, McBride GB. 2005. Estimation of nutrient sources and transport for New Zealand using the hybrid physical-statistical model SPARROW: *Journal of Hydrology* (New Zealand) 44: 1-27.

Figley W, Pyle B, Halgren B. 1979. Socioeconomic impacts. In: Oxygen depletion and associated benthic moralities in New York Bight, 1976. R.L. Swanson and C.J. Sindermann (eds.). *NOAA Professional Paper* 11, pp. 315-322.

Galloway JN, Townsend AR, Erisman JW, Bekunda M, Cai Z, Freney JR, Martinelli LA, Seitzinger SP, Sutton MA. 2008. Transformation of the nitrogen cycle: recent trends, questions, and potential solutions. *Science* 320: 889- 892.

Gamenick I, Jahn A, Vopel K, Giere O. 1996. Hypoxia and sulphide as structuringfactors in amacrozoobenthic community on the Baltic Sea shore: Colonisation studiesand tolerance experiments. *Mar. Ecol. Prog. Ser.* 144: 73–85.

Gardiner ES, Lockhart BR. 2007. Bottomland oak afforestation in the Lower Mississippi. *International Oak Journal* 18: 56-64.

Gardiner ES, Oliver JM. 2005. Restoration of bottomland hardwood forests in Lower Mississippi Alluvial Valley, USA. In Stanturf, J. A. and P. Madsen (eds) Restoration of boreal and temperate forests, USA Boca Raton, FL: CRC Press, pp. 235-251.

Garlo EV, Milstein CB, Jahn AE. 1979. Impact of hypoxic conditions in the vicinity of Little Egg Inlet, New Jersey in summer 1976. *Estuar. Coast. Mar. Sci.* 8: 421-432.

Glenn SM, Crowley MF, Haidvogel DB, Song YT. 1996. Underwater observatory captures coastal upwelling off New Jersey, Earth Space, 9, pp. 9–11.

Glenn S, Arnone R, Bergmann T, Bissett WP, Crowley M, Cullen J, Gryzmski J, Haidvogel D, Kohut J, Moline M, Oliver M, Orrico C, Sherrell R, Song T, Weidemann A, Chant R, Schofield O. 2004. Biogeochemical impact of summertime coastal upwelling on the New Jersey shelf. *J. Geophy. Res.* 109:C12S02.

Global Climate Change Impacts in the United States, Thomas R. Karl, Jerry M. Melillo, and Thomas C. Peterson, (eds.). Cambridge University Press, 2009.

Grantham BA, Chan F, Nielsen KJ, Fox DS, Barth JA, Huyer A, Lubchenco J, Menge BA. 2004. Upwelling-driven nearshore hypoxia signals ecosystem and oceanographic changes in the northeast Pacific. *Nature* 429: 749-754.

Green CH, Tomer MD, DiLuzio M, Arnold JG. 2006. Hydrologic evaluation of the soil and water assessment tool for a large tile-drained watershed in Iowa: *Transactions of the American Society of Agricultural and Biological Engineers 49*(2): 413–422.

Greene RM, Lehrter JC, Hagy JD III. 2009. Multiple regression models for hindcasting and forecasting midsummer hypoxia in the Gulf of Mexico. *Ecological Applications*: 19(5): 1161-1175.

Grizzetti B, Bouraoui F, De Marsily G. 2005. Modelling nitrogen pressure in river basins: A comparison between a statistical approach and the physically-based SWAT model, Physics and Chemistry of the Earth, *Assessment of Anthropogenic Impacts on Water Quality* 30(8-10): pp. 508-517.

Gulf of Mexico Hypoxia Monitoring Implementation Plan Steering Committee. 2009. Gulf of Mexico Hypoxia Monitoring Implementation Plan. An outcome from the Summit on Long-Term Monitoring of the Gulf of Mexico: *Developing the Implementation Plan for an Operational Observation System.* 37 pp.

Haas LW. 1977. Effect of spring-neap tidal cycle on vertical salinity structure of James, York and Rappahannock Rivers, Virginia, USA. E*stuarine and Coastal Marine Science* 5: 485-496.

Hagy JD, Boynton WR, Keefe CW, Wood KV. 2004. Hypoxia in Chesapeake Bay, 1950-2001: Long-term change in relation to nutrient loading and river flow. *Estuaries* 27(4): 634-658.

Hagy J D III, Lehrter JC, Murrell MC. 2006. Effects of Hurricane Ivan on water quality in Pensacola Bay, FL USA. *Estuaries and Coasts* 29(6A): 919-925.

Hagy JD III, Murrell MC. 2007. Susceptibility of a Gulf of Mexico estuary to hypoxia: An analysis using box models. *Estuarine, Coastal and Shelf Science* 74: 239-253.

Hagy JD III, Kurtz JC, Greene RM. 2008. An approach for developing numeric nutrient criteria for a Gulf coast estuary. US Environmental Protection Agency, Office of Research and Development, National Health and Environmental Effects Research Laboratory, Research Triangle Park, NC. EPA/600/R-08/004. 48 pp.

Hansen IS, Keul N, Sorensen JT, Erichsen A, Andersen JH. 2007. Oxygen maps for the Baltic Sea. BALANCE Interim Report No. 17

Harding LW Jr., Perry ES. 1997. Long-term increase of phytoplankton biomass in Chesapeake Bay, 1950–1994. *Marine Ecology Progress Series* 57: 39–52.

Hawley N, Johengen TH, Rao YR, Ruberg SA, Beletsky D, Ludsin SA, Eadie BJ, Schwab DJ, Croley TE, Brandt SB. 2006. Lake Erie Hypoxia Prompts Canada-US Study. *EOS Transactions* 87(32): 313-324.

Helly JJ, Levin LA. 2004. Global distribution of naturally occurring marine hypoxia on continental margins. *Deep-Sea Res.* (Part I) 51: 1159-1168.

Hetland RD, DiMarco SF. 2008. How does the character of oxygen demand control the structure of hypoxia on the Texas-Louisiana continental shelf? *Journal of Marine Systems* 70: 49-62.

Holman JD, Burnett KG, Burnett LE. 2004. Effects of hypercapnic hypoxia on the clearance of Vibrio campbellii in the Atlantic blue crab, Callinectes sapidus Rathbun. *Biol. Bull.* 206: 188–196.

Hong B, Swaney DP. 2007. Regional Nutrient Management (ReNuMa) Model, Version 1.0. User's Manual. http:// www.eeb.cornell.edu/ biogeo /nanc/ReNuMa/ReNuMa.zip.

Howarth RW, Billen G, Swaney D, Townsend A, Jaworski N, Lajtha K, Downing JA, Elmgren R, Caraco N, Jordan T,

Berendse F, Freney J, Kudeyarov V, Murdoch P, Zhu ZL. 1996. Regional nitrogen budgets and riverine N and P fluxes for the drainages to the North Atlantic Ocean: natural and human influences. *Biogeochemistry* 35: 75-139.

Hu X, McIssac GF, David MB, Louwers CAL. 2007. Modeling riverine nitrate export from an east-central Illinois watershed using SWAT: *Journal of Environmental Quality* 36: 996–1005.

Jaynes DB, James DE. 2007. Extent of Farm Drainage in the US. USDA Agricultural Research Service publication. http://www.ars.usda.gov/ SP2UserFiles/Place/36251500/ TheExtentofFarmDrainageintheUnitedStates.pdf.

Jha M, Arnold JG, Gassman PW, Giorgi F, Gu RR. 2006. Climate change sensitivity assessment on Upper Mississippi River Basin streamflows using SWAT: *Journal of American Water Resources Association* 42(4): 997–1016.

Justic′ D, Rabalais NN, Turner RE. 1996. Effects of climate change on hypoxia in coastal waters: a doubled CO2 scenario for the northern Golf of Mexico. *Limnol. Oceanogr.* 41: 992-1003.

Justic′ D, Rabalais NN, Turner RE. 2002. Modeling the impacts of decadal changes in riverine fluxes on coastal eutrophication near the Mississippi River Delta. *Ecological Modeling* 152: 33-46.

Justic′ D, Bierman VJ Jr, Scavia D, Hetland RD. 2007. Forecasting gulf's hypoxia: the next 50 years? Estuar. Coasts 30: 791-801.

Kansas Department of Health and Environment. 2004. Surface water nutrient reduction plan: http://www.kdhe.state. ks.us/water/download/ ks_nutrient_ reduction_plan_12_29_final.pdf (accessed March 1, 2005).

Kaplan RS, Norton DP. 2008. Mastering the management system. Harvard Business Review, January 2008 63-77.

Karlson K, Rosenberg R, Bonsdorff E. 2002. Temporal and spatial large-scale effects of eutrophication and oxygen deficiency on benthic fauna in Scandinavian and Baltic waters - A review. Oceanogr. *Mar. Biol. Ann. Rev.* 40: 427- 489.

Kemp WM, Boynton WR, Adolf JE, Boesch DF, Boicourt WC, Brush G, Cornwell JC, Fisher TR, Glibert PM, Hagy JD, Harding LW, Houde ED, Kimmel DG, Miller WD, Newell RIE, Roman MR, Smith EM, Stevenson JC. 2005. Eutrophication of Chesapeake Bay: historical trends and ecological interactions. *Mar. Ecol. Prog. Ser.* 303: 1-29.

Kennish MJ, Bricker SB, Dennison WC, Glibert PM, Livingston RJ, Moore KA, Noble RT, Paerl HW, Ramstack JM, Seitzinger S, Tomasko DA, Valiela I. 2007. Barnegat Bay–Little Egg Harbor estuary: case study of a highly eutrophic coastal bay system. *Ecological Applications* 17:S3–S16.

Kidwell D, Lewitus A, Jewett E, Brandt S, Mason D. 2009. Ecological Impacts of Hypoxia on Living Resources. *Journal of Experimental Marine Biology and Ecology* 381: S1-S3.

Koepfler E, Lake S, Smith EM, Bennett J, Libes S. 2007. Examination of Inner Shelf Water Quality in Long Bay, South Carolina, using Dataflow. Estuarine Research Federation Meeting Abstract. Providence Rhode Island, November 2007.

Koroncai R, Linker L, Sweeney J, Batuik R. 2003. Setting and allocating the Chesapeake Bay basin nutrient and sediment loads: the collaborative process, technical tools, and innovative approaches. *US Environmental Protection Agency*, Annapolis, MD.

Kramer DL. 1987. Dissolved oxygen and fish behaviour. *Environ. Bio. Fishes* 18: 81-92.

Landsberg JH, Flewelling LJ, Naar J. 2009. *Karenia brevis* red tides, brevetoxins in the food web, and impacts on natural resources: Decadal advancements. *Harmful Algae* 8(4): 598-607.

Langland MJ, Raffensperger JP, Moyer DL, Landwehr JM, Schwarz GE. 2006. Changes in Streamflow and Water Quality in Selected Nontidal Basins in the Chesapeake Bay Watershed, 1985-2004. Scientific Investigations Report 2006–5178; US Department of the Interior, US *Geological Survey.* 74 pp.

Le Moullac G, Soyez C, Salnier D, Ansquer D, Avarre JC, Levy P. 1998. Effect of hypoxic stress on the immune response and the resistance to vibriosis of the shrimp Penaeus stylirostris. *Fish Shellfish Immunol.* 8: 621–629.

Lee YJ, Lwiza KMM. 2008. Characteristics of bottom dissolved oxygen in Long Island Sound, New York. Estuar. *Coast. Shelf Sci.* 76: 187-200.

Leonard CL, McClintock JB. 1999. The population dynamics of the brittlestar Ophioderma brevispinum in near- and farshore seagrass habitats of Port Saint Joseph Bay, Florida. *Gulf of Mexico Science* 17: 87-94.

Lewis BL, Glazer BT, Montbriand PJ, Luther III GW, Nuzzio DB, Deering T, Ma S, Theberge S. 2007. Short-term and interannual variability of redox-sensitive chemical parameters in hypoxic/anoxic bottom waters of the Chesapeake Bay. *Marine Chemistry* 105 (3-4): 296-308.

Lewitus AJ, Kidwell DM, Jewett EB, Brandt S, Mason DM (eds.). 2009. Ecological Impacts of Hypoxia on Living Resources. *Journal of Experimental Marine Biology and Ecology*. Vol 381, Supplement 1, pp. S1-S216.

Li D, Zhang J, Huang D, Wu Y, Liang J. 2002. Oxygen depletion off the Changjiang (Yangtze River) Estuary. Science in China 45: 1137–1146.

Lipton D, Hicks R. 2003. The cost of stress: low dissolved oxygen and econimic benefits of recreational striped bass (*Morone saxatilis*) fishing in the Patuxent River. *Estuaries* 26: 310-315.

Ludsin S, Vanderploeg H, Pothoven S, Mason D, Hook T, Brandt S, Hawley N, Ruhberg S, Yerubundi R. 2009. Hypoxia effets on habitat and prey availability for randbow smelt in central Lake Erie. *Journal of Experimental Marine Biology and Ecology.*

Luettich RA, Jr., Carr SD, Reynolds-Fleming JV, Fulcher CW, McNinch JE. 2002. Semi-diurnal seiching in a shallow, microtidal lagoonal estuary. *Continental Shelf Research* 22: 1669-1681.

Luther GW, Ma SF, Trouwborst R, Glazer B, Blickley M, Scarborough RW, Mensinger MG. 2004. The roles of anoxia, H2S, and storm events in fish kills of dead-end canals of Delaware inland bays. *Estuaries* 27: 551-560.

Maresch W, Walbridge MR, Kugler D. 2008. Enhancing conservation on agricultural landscapes*: A new direction for the Conservation Effects Assessment Project.* 63 (6): 198A – 203A.

Marcus NH, Richmond C, Sedlacek C, Miller GA, Oppert C. 2004. Impact of hypoxia on the survival, egg production and population dynamics of Acartia tonsa Dana. *Journal of Experimental Marine Biology and Ecology* 301(2): 111- 128.

Mason WT, Jr. 1998. Macrobenthic monitoring in the Lower St. Johns River, *Florida. Environ. Monit. Assess.* 50: 101- 130.

May E. 1973. Extensive oxygen depletion in Mobile Bay, Alabama. Limnol. Oceanogr. 18: 353-366.

McMahon G, Alexander RB, Qian S. 2003. Support of TMDL programs using spatially referenced regression models: *ASCE Journal of Water Resources Planning and Management*, v. 129, pp. 315-329.

Mee LD. 1992. The Black Sea in crisis: a need for concerted international action. Ambio 21: 278-286. Mee L. 2006. Reviving dead zones. *Scientific American* 295: 78-85.

Melrose DC, Oviatt CA, Berman MS. 2007. Hypoxic events in Narragansett Bay, Rhode Island, during the summer of 2001. *Estuar. Coasts* 30: 47-53.

Miller JR, Russell GL. 1992. The impact of global warming on river runoff. *J. Geophy. Res.* 97: 2757-2764.

Mississippi River/Gulf of Mexico Watershed Nutrient Task Force. 2001. Action plan for reducing, mitigating, and controlling hypoxia in the northern Gulf of Mexico: Washington, DC.

Mississippi River/Gulf of Mexico Watershed Nutrient Task Force. 2004. Science Strategy to Support Management Decisions Related to Hypoxia in the Northern Gulf of Mexico and Excess nutrients in the Mississippi River Basin.

Mississippi River/Gulf of Mexico Watershed Nutrient Task Force. 2008. Gulf Hypoxia Action Plan 2008 for Reducing, Mitigating and Controlling Hypoxia in the Northern Gulf of Mexico and Improving Water Quality in the Mississippi River Basin. Washington DC.

Montagna PA, Ritter C. 2006. Direct and indirect effects of hypoxia on benthos in Corpus Christi Bay, Texas, USA. *J. Exper. Mar. Biol. Ecol.* 330: 119-131.

Moore RB, Johnston CM, Robinson KW, Deacon JR. 2004. Estimation of total nitrogen and phosphorus in New England streams using spatially referenced regression models: US Geological Survey Scientific Investigations Report 2004-5012, 50 p. http://water.usgs.gov/pubs/sir/2004/5012/.

Najjar RG, Walker HA, Anderson PJ, Barron EJ, Bord RJ, Gibson JR, Kennedy VS, Knight CG, Megonigal JP,

O'Connor RE, Polsky CD, Psuty NP, Richards BA, Sorenson LJ, Steele EM, Swanson RS. 2000. The potential impacts of climate change on the mid-Atlantic coastal region. *Climate Res.* 14: 219-233.

NASS (National Agriculture Statistics Service). 2007. Acreage June 2007. U. S. Department of Agriculture, Agriculture Statistics Board, Washington DC.

NASS (National Agriculture Statistics Service). 2008. Acreage June 2008. U. S. Department of Agriculture, Agriculture Statistics Board, Washington DC.

Neitsch SL, Arnold JG, Kiniry JR, Srinivasan R, Williams JR. 2004. Soil and water assessment tool input/output file documentation, version 2005: Temple, TX, US Department of Agriculture, Agricultural Research Service, Grassland, Soil and Water Research Laboratory, available online at: ftp://ftp.brc.tamus.edu.pub/outgoing/ sammons/swat2005 (accessed 11/28/ 06).

NEIWPCC (New England Interstate Water Pollution Control Commission). 2004. New England SPARROW water quality model: Interstate Water Report, v. 1, no. 3, p. 6-7. http://www.neiwpcc.org/PDF_Docs/iwr_s04.pdf

Niklitschek E, Secor DH. 2005. Modeling spatial and temporal variation of suitable nursery habitats for Atlantic sturgeon in the Chesapeake Bay. Est. Coast. *Shelf Science* 64, 135-148.

Nixon SW. 1995. Coastal marine eutrophication: A definition, social causes, and futurre concerns. *Ophelia* 41: 199- 219.

Nixon SW, Buckley B, Granger S, Harris LA, Oczkowski AJ, Fulweiler RW, Cole LW. 2008. Nutrient (N and P) inputs to Narragansett Bay: Past, present, and future. In Ecosystem Based Management: A Case Study of Narragansett Bay. *Springer Series in Environmental Management*, eds. A. Desbonnet, and B. A. Costa-Pierce, 101–176. New York: Springer.

NOAA. 2004. Lake Erie Research Planning Workshop Report. National Oceanic and Atmospheric Administration, *Great Lakes Environmental Research Laboratory*. Ann Arbor, MI. 28 pp.

NRC (National Research Council). 2000. Clean Coastal Waters: Understanding and Reducing the Effects of Nutrient Pollution. National Academy Press Washington, DC.

NRC (National Research Council). 2007. Water Implications of Biofuels Production in the United States. National Academy Press Washington, DC. 86 pp.

NRC (National Research Council). 2008. Nutrient Control Actions for Improving Water Quality in the Mississippi River Basin and Northern Gulf of Mexico. The National Academies Press, Washington DC. 66 pp.

O'Connor T, Whitall D. 2007. Linking hypoxia to shrimp catch in the northern Gulf of Mexico. *Marine Pollution Bulletin* 54(4): 460-463.

Park K, Kim CK, Schroeder WW. 2007. Temporal variability in summertime bottom hypoxia in shallow areas of Mobile Bay, Alabama. *Estuaries and Coasts* 30(1): 54-65.

Patrick R. 1988. Changes in the chemical and biological characteristics of the Upper Delaware River estuary in response to environmental laws. In: Majumdar E, Miller W, Sage LE (eds.), pp. 332-359. Pennsylvania Academy of Science, Philadelphia, PA.

Potter SR, Andrews S, Atwood JD, Kellogg RL, Lemunyon J, Norfleet L, Oman D. 2006. Model simulation of soil loss, nutrient loss, and change in organic carbon associated with crop production. USDA Natural Resource Conservation Service: http://www.nrcs. usda.gov/technical/nri/ceap/ crop landreport/.

Pouyat RV, Pataki DE, Belt KT, Groffman PM, Hom J, Band LE. 2007. Effects of land-use change on biogeochemical cycles. In: Canadell JG, Pataki DE, Pitelka LF (eds.) Terrestrial ecosystems in a changing world. Berlin. *SpringerVerlag*. pp. 45-58.

Preston SD, Brakebill JW. 1999. Application of spatially referenced regression modeling for the evaluation of total nitrogen loading in the Chesapeake Bay watershed: Baltimore, MD, US Geological Survey Water Resources Investigations Report 99-4054, 11 pp., http://md.water.usgs.gov /publi cations/wrir-99-4054/.

Rabalais NN, Harper DE, Turner RE. 2001. Responses of nekton and demersal and benthic fauna to decreasing oxygen concentrations. pp. 115-128. In: Rabalais N.N., Turner R.E. (eds.) Coastal Hypoxia Consequences for Living *Resources and Ecosystems*. American Geophysical Union, Washington, D.C.

Rabalais NN, Turner RE. 2001. "Coastal hypoxia. Consequences for living resources and ecosystems". Coastal and Estuarine Studies, No. 58, American Geophysical Union, Washington DC. 463 pp.

Rabalais NN, Turner RE, Sen Gupta BK, Boesch DF, Chapman P, Murrell MC. 2007. Hypoxia in the northern Gulf of Mexico: Does the science support the plan to reduce, mitigate and control hypoxia? *Estuaries and Coasts*. 30(5): 753-772.

Rabalais NN, Turner RE, Díaz RJ, D. Justić. 2009. Global Change and Eutrophication of Coastal Waters. *ICES Journal of Marine Science*, 66: 000–00.

Reynolds-Fleming JV, Luettich RA Jr. 2004. Wind-driven lateral variability in a partially mixed estuary. Estuarine, *Coastal and Shelf Science*. 60: 395-407.

Ribaudo MO, Heimlich R, Claassen R, Peters M. 2001. Least-cost management of nonpoint source pollution: source reduction versus interception strategies for controlling nitrogen loss in the Mississippi Basin, *Ecological Economics*. 37: 183–197.

Ribaudo MO, Heimlich R, Peters M. 2005. Nitrogen sources and Gulf hypoxia: potential for environmental credit trading, *Ecological Economics*. 52(2): 159-168.

Ritter C, Montagna PA. 1999. Seasonal hypoxia and models of benthic response in a Texas bay. *Estuaries* 22: 7-20.

Runkel RL, Crawford CG, Cohn TA. 2004. Load estimator (LOADEST) - A FORTRAN program for estimating constituent loads in streams and rivers. US Geological Survey Techniques and Methods, book 4, chap. A5, 69 pp. http://pubs.usgs.gov/tm/2005/tm4A5/.

Sanger DM, Arendt MD, Chen Y, Wenner EL, Holland AF, Edwards D, Caffrey J. 2002. A Synthesis of Water Quality Data: National Estuarine Research Reserve System-wide Monitoring Program (1995-2000). National Estuarine Research Reserve Technical Report Series 2002: 3, South Carolina Department of Natural Resources, Marine Resources Division Contribution No. 500, Charleston, SC.

Sagasti A, Schaffner LC, Duffy JE. 2001. Effects of periodic hypoxia on mortality, feeding and predation in an estuarine epifaunal community. *J. Exp. Mar. Biol. Ecol.* 258: 257 283.

Santhi C, Arnold JG, Williams JR, Dugas WA, Srinivasan R, Hauck LM. 2001. Validation of the SWAT model on a large river basin with point and nonpoint sources: *Journal of the American Water Resources Association* 37(5): 1169–1188.

Scavia D, Rabalais NN, Turner RE, Justic D, Wiseman, WJ, Jr. 2003. Predicting the response of Gulf of Mexico hypoxia to variations in Mississippi River nitrogen load. *Limnology and Oceanography* 48: 951–956.

Scavia D, Justic D, Bierman VJ. 2004. Reducing hypoxia in the Gulf of Mexico: advice from three models. *Estuaries* 27: 419–425.

Schulte LA, Liebman M, Asbjornsen H, Crow TR. 2006. Agroecosystem restoration through strategic integration of perennials. *Journal of Soil and Water Conservation* 61: 165A-169A.

Seitz RD, Dauer DM, Llansó RJ, Long WC. 2009 Broad-scale effects of hypoxia on benthic community structure in Chesapeake Bay, USA. . *Journal of Experimental Marine Biology and Ecology*.

Shen J, Wang T, Herman J, Mason P, Arnold GL. 2008. Hypoxia in a coastal embayment of the Chesapeake Bay: A model diagnostic study of oxygen dynamics. *Estuaries and Coasts* 31: 652-663.

Shchepetkin AF, McWilliams JC. 2005. The regional oceanic modeling system (ROMS): a split-explicit, free-surface, topographically-following-coordinate oceanic model. *Ocean Modelling* 9, 347–404.

Shen J, Wang T, Herman J, Mason P, Arnold GL. 2008. Hypoxia in a Coastal Embayment of the Chesapeake Bay: A Model Diagnostic Study of Oxygen Dynamics, *Estuaries and Coasts*. 31(4): 652-663.

Sindermann C, Swanson R. 1980. Chapter 1. Historical and Regional Perspective. In: Oxygen depletion and associated benthic mortalities in New York Bight, 1976. R.L. Swanson and C.J. Sindermann (eds.) Rockville. US Department of Commerce National Oceanic and Atmospheric Administration. pp.1-16.

Smith RA, Alexander RB, Wolman MG. 1987. Water-quality trends in the nation's rivers. Science 235: 1607-1615.

Smith RA, Schwarz GE, Alexander RB. 1997. Regional interpretation of water-quality monitoring data: *Water Resources Research* 33: 2781-2798.

Stow CA, Qian SS, Craig JK. 2005. Declining threshold for hypoxia in the Gulf of Mexico. *Environmental Science and Technology* 39: 716-723.

Strosneider WH, Hitchcock DR, Burke MK, Lewitus AJ. 2007. Predicting hydrology in wetlands designed for coastal stormwater management. American Society of Agricultural and Biological Engineers. Paper Number: 077084. An ASABE Meeting Presentation. pp. 1-17.

Tanner CA, Burnett LE, Burnett KG. 2006. The effects of hypoxia and pH on phenoloxidase activity in the Atlantic blue crab, Callinectes sapidus. *Comp. Biochem. Physiol.* 144: 218-33.

Taylor JC, Miller JM. 2001. Physiological performance of juvenile southern flounder, *Paralichthys lethostigma* (Jordan and Gilber, 1884), in chronic and episodic hypoxia. *Journal of Experimental Marine Biology and Ecology* 258: 195-214.

Taylor JC, Rand PS, Jenkins J. 2007. Swimming behavior of juvenile anchovies (*Anchoa* spp.) in an episodically hypoxic estuary: implications for individual energetics and trophic dynamics. *Mar. Biol.* 152: 939–57.

Tenore KR. 1972. Macrobenthos of the Pamlico River estuary, North Carolina. Ecol. Monogr. 42: 51-69. Thronson A, Quigg A. 2008. Fifty-five years of fish kills in coastal Texas. *Estuar. Coasts* 31: 802–813.

Thurston RV. 2002. Fish physiology, toxicology, and water quality proceedings of the sixth international symposium. La Paz, B.C.S. Mexico: Ecosystems Research Division.

Tilman D, Fargione J, Wolff B, D'Antonio C, Dobson A, Howarth R, Schindler D, Schlesinger WH, Simberloff D, Swackhamer D. 2001. Forecasting Agriculturally Driven Global Environmental Change. *Science* 292: 281-284.

Thomas P Md, Rahman S, Khan IA, Kummer JA. 2007. Widespread endocrine disruption and reproductive impairment in an estuarine fish population exposed to seasonal hypoxia. *Proceedings of the Royal Society* B 274(1626): 2693–2702.

Turner RE, Rabalais NN, Fry B, Atilla N, Milan CS, Lee JM, Normandeau C, Oswald TA, Swenson EM, Tomasko DA. 2006a. Paleo-indicators and water quality change in the Charlotte Harbor estuary (Florida). *Limnol. Oceanogr.* 51: 518–533.

Turner RE, Rabalais NN, Justic′ D. 2006b. Predicting summer hypoxia in the northern Gulf of Mexico: Riverine N, P, and Si loading. *Marine Pollution Bulletin* 52: 139-148.

Turner RE, Rabalais NN, Justic D. 2008. Gulf of Mexico Hypoxia : Alternate States and a Legacy. *Environ. Sci. Tech.* 42: 2323 – 2327.

Tyler RM, Targett TE. 2007. Juvenile weakfish Cynoscion regalis distribution in relation to diel-cycling dissolved oxygen in an estuarine tributary. *Mar. Ecol. Prog. Ser.* 333: 257-269.

Tyler RM, Brady DC, Targett TE. 2008. Temporal and spatial dynamics of diel-cycling hypoxia in estuarine tributaries. *Estuaries and Coasts* 32: 123-145.

US Commission on Ocean Policy. 2004. An Ocean Blueprint for the 21st Century. Final Report. Washington, DC, 2004. ISBN#0–9759462–0–X.

US EPA. 2002. National Water Quality Inventory 2000 Report. Office of Science and Technology/Office of Water, Washington, DC.

US EPA. 2003. Biological Evaluation for the Issuance of Ambient Water Quality Criteria for Dissolved Oxygen, Water Clarity and Chlorophyll a for the Chesapeake Bay and Its Tidal Tributaries. Region III, Chesapeake Bay Program, Annapolis, Maryland. 61 pp.

US EPA. 2005. National Coastal Condition Report II. Office of Research and Development/Office of Water, Washington, DC. US EPA 2002.

US EPA. 2008. National Coastal Condition Report III. Office of Research and Development/Office of Water, Washington, DC.

US EPA 2007. Hypoxia in the Northern Gulf of Mexico: An Update by the EPA Science Advisory Board. EPA-SAB08-004. EPA Science Advisory Board, Washington DC.

US EPA. 2008. National coastal condition report III. EPA/842-R-08-002. Office of Research and Development/Office of Water, Washington, DC. 300 pp.

Vache KB, Eilers JM, Santelman MV. 2002. Water quality modeling of alternative agricultural scenarios in the US corn belt: *Journal of the American Water Resources Association* 38(2): 773–787.

Valiela I, Cole ML. 2002 Comparative evidence that salt marshes and mangroves may protect seagrass meadows from land-derived nitrogen loads. *Ecosystems* 5: 92-102.

Vaquer-Sunyer R, Duarte C. 2008. Thresholds of hypoxia for marine diversity. PNAS 105 (40): 15452–15457.

Verity PG, Alber M, Bricker SB. 2006. Development of hypoxia in well-mixed subtropical estuaries in the southeastern USA. *Estuar. Coasts* 29: 665-673.

Vitousek PM., Aber J, Howarth RW, Likens GE, Matson PA, Schindler DW, Schlesinger WH, Tilman GD. 1997. Human Alteration of the Global Nitrogen Cycle: Causes and Consequences. *Issues in Ecology*, 1: 1-16.

Ward MH, deKok TM, Levallois P. 2005. Workgroup report: drinking-water nitrate and health-recent findings and research needs. *Environmental Health Perspectives* 113: 1607-1614.

Whitledge TE. 1985. Nationwide review of oxygen depletion and eutrophication in estuarine and coastal waters. Executive Summary. Report to US Dept. of Commerce, NOAA, National Ocean Service. Rockville, MD. 28 p.

Whitmore CM, Warren CE, Doudoroff P. 1960. Avoidance reactions of salmonid and centrarchid fishes to low oxygen concentrations. *Trans. Am. Fisher. Soc.* 89: 17–26.

Wu RSS, Zhou BS, Randall DJ, Woo NYS, Lam PKS. 2003. Aquatic hypoxia is an endocrine disruptor and impairs fish reproduction. *Environ. Sci. Tech.* 37: 1137-1141.

Zhang H, Ludsin SA, Roman MR, Boicourt WC, Zhang X, Kimmel DG, Adamack AT, Mason DM, Brandt SB. 2009. Hypoxia-driven changes in the behavior and spatial distribution of pelagic fish and zooplankton in the northern Gulf of Mexico. *Journal of Experimental Marine Biology and Ecology* 381: S 80 - 91.

Appendix I: Federal Agency Hypoxia or Hypoxia-related Research

I.A. Department of Agriculture

I.A.1. Agricultural Research Service

The *Agricultural Research Service (ARS)* is an in-house scientific research agency of the US Department of Agriculture.

ARS conducts research to develop and transfer solutions to agricultural problems of high national priority and disseminates information to: 1) ensure high-quality, safe food, and other agricultural products; 2) assess the nutritional needs of Americans; 3) sustain a competitive agricultural economy; 4) enhance the natural resource base and the environment; and 5) provide economic opportunities for rural citizens, communities, and society as a whole.

The *Water Availability and Watershed Management National Program* is part of Natural Resources and Sustainable Agricultural Systems, one of four broad areas for ARS research. The national program structure allows ARS to link USDA scientists in laboratories around the country to address research problems of local, regional, national, and international interest, such as hypoxia. http://www.ars.usda.gov/ research/programs/ programs.htm?NP_CODE=211.

The *Conservation Effects Assessment Project (CEAP)* provides the farming community, the conservation community, the general public, the Office of Management and Budget, legislators, and others involved with environmental policy issues, with an account of the environmental effects or benefits obtained from USDA conservation program expenditures.

CEAP is jointly managed by ARS; the Cooperative State Research, Education, and Extension Service (CSREES); Farm Service Agency (FSA); and Natural Resources

Conservation Service (NRCS). http://www.nrcs. usda.gov/ TECHNICAL/NRI/ceap/ *STEWARDS (Sustaining the Earths Watersheds Agricultural Research Data System)* was created to house data collected as part of CEAP assessment for croplands.

This data system organizes and documents soil, water, climate, land-management, and socioeconomic data from multiple agricultural watersheds across the United States, allowing users to search, download, visualize, and explore data for research and conservation management purposes.

The National Agricultural Library, Beltsville, MD, has developed a series of six CEAP-related bibliographic databases that are available to the public through their website: http://www.nal.usda.gov/wqic/. This dynamic relational database offers more than 5,200 citations of current research findings from around the globe to help scientists and the public understand the environmental effects and benefits of conservation practices implemented through various USDA conservation programs. http://arsagsoftware.ars. usda.gov/stewards/.

I.A.2. Cooperative State Research, Education, and Extension Service

CSREES provides Federal assistance in the form of grants for research, education, and extension activities related to agriculture. The CSREES Water Program focuses on creating and disseminating knowledge that ensures a safe and reliable source of water to meet the needs for food, fiber, and energy production; human use and economic growth; and maintenance and protection of natural environmental systems and ecosystem services. CSREES' unique niche is conducting research, education, and extension programs to protect and improve water resources in agricultural, rural, and urbanizing watersheds (including forest lands, rangelands, and croplands). CSREES has been a partner with ARS, NRCS, and FSA in the CEAP. Thirteen watershed projects were jointly funded by CSREES and NRCS to evaluate the effectiveness of conservation practices on water quality at the watershed scale. These 13 watershed projects serve as examples of collaborative work between land grant universities and NRCS. The projects are unique in that they combine evaluation of the biophysical effects of conservation practices and the socioeconomic context of the watershed location. The watershed projects also combine research and extension/outreach activities – involving agricultural producers in project outcomes. Through CEAP, the land grant university system has developed substantial capacity to increase the understanding of effects of conservation practices and the effectiveness of conservation programs. CSREES and NRCS collaborated to fund two projects to synthesize results from the 13 watershed-scale projects. These two projects will build a knowledge base to evaluate impacts of conservation practices over broad regions, improve management of agricultural landscapes, and inform policy decisions at the local, state, and national scale. The two projects differ in their approach – the first uses a synoptic overview of the 13 watershed studies and the second uses a modeling framework to spatially distribute results from the 13 watershed to broader geographic regions. CSREES also provides funding for a national network of regional projects that focus on performance-based programs to address water resource issues. The ten regional projects are geographically aligned with US EPA regions (see www.usawaterquality.org). These regional projects are directly involved with addressing hypoxia through research, education, and extension programs on animal manure management, nutrient management, and social and economic policy sciences. They make research, education, and extension resources of the university system more accessible to Federal, State, and local water resources improvement efforts. Through these projects, new or expanded opportunities develop for agricultural producers and agriculturally impacted communities to adopt voluntary approaches for the improvement of water resources. They promote delivery of multi-state programming to address regional and/or multiregional water quality concerns.

I.A.3. Farm Service Agency

The primary **FSA** activity related to hypoxia is the Conservation Reserve Program (CRP). CRP is a voluntary program available to agricultural producers to help them safeguard environmentally sensitive land. Producers enrolled in CRP plant long-term, resource-conserving covers to improve the quality of water, control soil erosion, and enhance wildlife habitat. In return, FSA provides participants with rental payments and cost-share assistance. Contract duration is between 10 and 15 years.

CRP protects millions of acres of American topsoil from erosion and is designed to safeguard the Nation's natural resources. By reducing water runoff and sedimentation, CRP protects groundwater and helps improve the condition of freshwater and marine resources. Acreage enrolled in the CRP is planted with resource-conserving vegetative covers, making the program a major contributor to increased wildlife populations in many parts of the country. http://www.fsa.usda.gov/FSA/webapp?area=home&subject=cop r&topic=crp.

I.A.4. US Forest Service

The USDA Forest Service (http://www.fs.fed.us/research/) has a mission to sustain the health, diversity, and productivity of the Nation's forests and grasslands to meet the needs of present and future generations. The research and development arm of the Forest Service works at the forefront of science to improve the health and use of our Nation's forests and grasslands. Today, more than 500 Forest Service researchers work across the 50 states, US territories, and commonwealths in a range of biological, physical, and social science fields to promote sustainable management of the Nation's diverse forests and rangelands and the ecosystem services they provide. The work has a steady focus on informing policy and land management decisions, including those on degraded river ecosystems. The researchers work independently and with a range of partners, including other agencies, academia, nonprofit groups, and industry. The information and technology produced through basic and applied science programs is available to the public for its benefit and use.

With over two-thirds of the Nation's freshwater originating on forested lands, the role of the Forest Service to manage and restore watersheds is indispensable.

The Forest Service works to improve management practices and techniques, develop best management practices that are cost-effective, and lead efforts in the advancement of markets and payments for ecosystem services. In addition, the Forest Service is working to develop methods for assessing changes in watershed condition as a result of changing demographics and development pressures.

The Forest Service has developed a national strategy for consistent characterization of watershed condition. The characterization will be used to prioritize, identify and implement integrated watershed-scale improvement projects on National Forests. Phase 1 will determine the initial (baseline) watershed condition on National Forest System lands using consistent, broadly available data. Phase 2 is a six-step process for classifying watersheds at a finer scale and prioritizing them for treatment, identifying integrated projects that would move them into improved condition classes. Phase 3 involves validating and monitoring watershed condition.

I.A.5. Economic Research Service

The *Economic Research Service* (ERS) is the main source of economic information and research from USDA. It brings the perspective of economic analysis to critical issues confronting farmers, agribusiness, consumers, and policy makers.

In general, ERS's data program collects information that can be used in an assessment of policy options in the Mississippi Basin. In conjunction with USDA National Agricultural Statistics Service, ERS collects data through the Agricultural Resource Management Survey on production practices, conservation practices and input use for major crops in major crop producing states.

ERS uses these and other data to routinely report on the status and trends of water resource indicators such as tillage practices, conservation practices, chemical input use, and conservation program expenditures for water quality and water conservation. http://www.ers. usda.gov/.

I.B. Department of Commerce

I.B.1. National Oceanic and Atmospheric Administration

I.B.1.a. National Ocean Service

National Centers for Coastal Ocean Science, Center for Sponsored Coastal Ocean Research

Gulf of Mexico Ecosystems and Hypoxia Assessment Program (NGOMEX). To address the issue of hypoxia in the northern Gulf of Mexico, NOAA's Center for Sponsored Coastal Ocean Research supports multi-year, competitive, interdisciplinary research projects to develop a fundamental understanding of the northern Gulf of Mexico ecosystem with a focus on the causes and effects of the hypoxic zone and the prediction of its future extent and impacts.

The research program, which began in 2003 as a result of the HABHRCA 1998 legislation, focuses on developing a predictive capability for the Louisiana continental shelf ecosystem within an adaptive management framework by connecting monitoring, data analysis, model predictions, and management actions with continuous feedback for improvement. Current studies are documenting the dynamics of the hypoxic zone over the Louisiana continental shelf and helping to better define the biological, chemical, and physical processes that influence hypoxic zone development and determine its extent, and impacts on fisheries. http://www.cop. noaa.gov/stressors/pollution/ current/ gomex-factsheet.html.

Coastal Hypoxia Research Program (CHRP). The purpose of the CHRP program, which awarded its first research grants in 2005, is to expand NOAA's research capability to address hypoxia in other regions experiencing hypoxia beyond the northern Gulf of Mexico.

This program is providing research results and modeling tools to help coastal resource managers assess alternative management strategies and make informed decisions for preventing or mitigating the impacts of hypoxia on coastal ecosystems.

Determining the causes of hypoxia, developing the capability to predict its occurrence in response to varying levels of anthropogenic stress, and evaluating the subsequent ecological, economic, and social impacts are necessary to assess potential management alternatives.

Key areas of CHRP research include:

1) the development of a predictive capability for the spatial and temporal extent of hypoxia given current and potential anthropogenic and natural forcing scenarios and potential management alternatives,
2) the determination of the current ecological and economic impacts of hypoxia in a region,
3) the development of ecological forecasts and economic valuations of the impacts of changes in a hypoxic region's spatial or temporal extent, and
4) the development of models that predict the susceptibility of coastal systems to the formation of hypoxic waters, thereby allowing managers to better focus monitoring and assessment programs which should result in more efficient and successful protection and restoration efforts. http://www.cop.noaa.gov/stressors/pollution/current/gomex-factsheet. Html.

National Centers for Coastal Ocean Science, Center for Coastal Monitoring and Assessment.

NOAA's Center for Coastal Monitoring and Assessment recently completed an ambitious update to the National Estuarine Eutrophication Assessment (http://ccma.nos.noaa.gov/publications/eutroupdate), which assesses nutrient conditions and direct/indirect impacts in US estuaries. The report thoroughly analyzed available data and assigned condition rankings to estuaries.

Geographic accounts of hypoxia are included in the report. Changes in eutrophication status since the last assessment give a sense of the growing incidences of hypoxia in coastal waters.

Office of Ocean and Coastal Resource Management

National Estuarine Research Reserve System (NERRS). The principal mission of the NERRS System-Wide Monitoring Program (SWMP) is to develop quantitative measurements of short-term variability and long-term changes in the water quality, biotic diversity, and land-use/land -cover characteristics of estuaries and estuarine ecosystems for the purposes of contributing to effective coastal zone management.

Long-term monitoring and iterative habitat assessments conducted within the network of 27 NERRS sites are intended to improve the fundamental understanding of the temporal and spatial dynamics of estuarine processes, and to provide baseline information to evaluate subsequent changes in the ecological status of estuarine ecosystems in response to natural perturbations and anthropogenic disturbance.

Federal coordination of the NERRS SWMP allows for the recognition of significant ecological trends (improvement or degradation) in estuaries at national, regional, and local levels.

Moreover, continued operation of the long-term monitoring effort within the network of representative NERRS sites will provide valuable data to inform assessments and/or models on the cumulative effects of environmental stressors in estuarine ecosystems located throughout the US coastal zone. Monitoring at each SWMP site includes continuous measurements of dissolved oxygen levels and monthly nutrient sampling.

Between 2005 and 2007, NERRS organized the SWMP data into several regional reports, including for the Northeast, Mid-Atlantic and the Pacific coast.

I.B.1.b. Oceanic and Atmospheric Research

Great Lakes Environmental Research Laboratory

NOAA's Great Lakes Environmental Research Laboratory, in collaboration with researchers from the United States and Canada, is undertaking one of the largest, most comprehensive Lake Erie research field programs ever.

The International Field Years on Lake Erie (IFYLE) program, which started in 2005 and is projected to continue through 2010, aims to improve understanding of large-scale (time and space) driving forces of ecosystem dynamics.

The project has a special research focus on hypoxia and harmful algal blooms and their impacts in Lake Erie.

National Sea Grant College Program

The National Sea Grant College Program is a competitive awarded research program which leverages state participation with Federal funds. It has focused some research funding on hypoxia research, specifically in the Southeast coastal states.

I.B.1.c. National Marine Fisheries Service

NOAA National Marine Fisheries Service (NMFS) conducts fish surveys/mapping of the entire Gulf coast, with bottom dissolved oxygen as one of the measured parameters, in late June/early July and then again in early September as part of its routine Southeast Area Monitoring and Assessment Program's (SEAMAP's) groundfish surveys.

The dissolved oxygen data are provided to the public via the Gulf of Mexico Hypoxia Watch website (http://ecowatch.ncddc.noaa.gov/hypoxia), which is managed by NOAA's National Environmental Satellite, Data, and Information Service.

NOAA NMFS scientists have also studied the impact of hypoxia on Gulf coast fisheries. NMFS has also added dissolved oxygen monitoring to its salmonid surveys along the Pacific Northwest to monitor hypoxia on the continental shelf.

I.C. Department of Defense

I.C.1. US Army Corps of Engineers

System-Wide Water Resources Program (SWWRP)

The US Army Corps of Engineers (USACE) is developing watershed, regional, and system-wide assessment tools that are applicable to nutrient and sediment assessments and decision support. As part of the SWWRP, one-, two, and three-dimensional hydrologic, hydrodynamic, and water quality models have been developed and demonstrated in various watershed, riverine, reservoir, and estuarine applications.

Example applications include assessment of nutrient and sediment loading in the upper Eau Galle, Wisconsin, watershed and hydrodynamic assessments in the Upper Auglaize, Ohio, watershed. Basic research in nitrogen dynamics has been done in conjunction with pool fluctuations in the Upper Mississippi River.

Response in reservoir water quality to loadings has been studied in numerous USACE reservoirs. Response to nutrient and sediment loads in estuaries has been assessed in the Chesapeake Bay. Data management and visualization tools using CorpsGlobe (developed with Google Earth technologies) have been fielded for displaying multidimensional modeling results in a geospatial and temporal context for decision support.

Chesapeake Bay Environmental Model Package (CBEMP)

The CBEMP is the product of a twenty-year partnership among the USACE Baltimore District, the USACE Engineer Research and Development Center, and the EPA Chesapeake Bay Program.

The CBEMP is a coupled system of models aimed and managing and reducing eutrophication in the Bay. In particular, the model is aimed at restoring living resources, reducing anoxia, and improving water clarity.

The present phase of the study continues the partnership and is focused on the simulation of suspended solids and living resources interactions. The current focus is to assess how management of suspended solids in the watershed may improve Chesapeake Bay water quality and remove water quality impairments, such as low dissolved oxygen, algal blooms, and poor water clarity.

USACE has completed an initial calibration with runs of various management scenarios and is revising the model based on the results of the initial scenarios.

I.D. US Environmental Protection Agency

I.D.1. Regional Offices

Great Lakes National Program Office

The EPA Great Lakes National Program Office (GLNPO) brings together Federal, state, tribal, local, and industry partners in an integrated, ecosystem approach to protect, maintain, and restore the chemical, biological, and physical integrity of the Great Lakes.

In respect to hypoxia, the program monitors Lake ecosystem indicators, such as dissolved oxygen, and manages and provides public access to Great Lakes data (http://www.epa.gov/glnpo/monitor.html). Each year, GLNPO uses its funding to assist Great Lakes partners in these areas through grants, interagency agreements, and contracts (http://www.epa.gov/glnpo/fund/glf.html). In 2002, the GLNPO held a funding competition focused specifically on hypoxia in western Lake Erie.

Gulf of Mexico Program Office

The Gulf of Mexico Program was formed in 1988 by the EPA as a nonregulatory, inclusive partnership to provide a broad geographic focus on the major environmental issues in the Gulf. The mission of the Program is "to facilitate collaborative actions to protect, maintain, and restore the health and productivity of the Gulf of Mexico in ways consistent with the economic well-being of the Region." The partnership includes representatives from state and local governments and citizens in each of the five Gulf States; the private sector (business and industry); Federal agencies responsible for research, monitoring, environmental protection, and natural resource management; and the academic community.

This partnership has been very effective in providing a mechanism for addressing complex problems that cross Federal, state, and international jurisdictional lines; a better coordination among Federal, state, and local programs that has increased the effectiveness and efficiency of the long-term commitment to manage and protect Gulf resources; a regional perspective to access and provide the information and address research needs required for effective management decisions; and a forum for affected groups using the Gulf, public and private educational institutions, and the general public to participate in the solution process.

Through its partnerships, the Program is working with the scientific community; policymakers at the Federal, state, and local levels; and the public to help preserve and protect the Gulf. It has made significant progress in identifying the environmental issues in the Gulf ecosystem and in organizing a program to address those issues.

The Program provides a tool to leverage the resources of 18 different Federal agencies, a variety of environmentally minded agencies from the states of Alabama, Florida, Louisiana, Mississippi, and Texas; and numerous public and private organizations. Spanning the broad range of environmental concerns from the local to the regional and national presents a significant challenge to the agencies and organizations that have partnered under the banner of the Gulf of Mexico Program, and the Program has met that challenge.

Chesapeake Bay Program Office

The Chesapeake Bay Program Office represents the Federal government in the implementation of strategies to meet the restoration goals of the Chesapeake Bay Program.

The Chesapeake Bay Program is a unique regional partnership leading and directing restoration of the Chesapeake Bay. Its mission is to lead and empower others to protect and restore the Chesapeake Bay ecosystem for future generations.

I.D.2. Office of Water

Office of Wetlands, Oceans, and Watersheds

The office of Wetlands, Oceans, and Watersheds (OWOW) provides leadership, policy directions, and financial support to ten regional offices as well as states, tribes, and territories. OWOW administers a number of programs that are directed at reducing nutrient pollution and its resultant impairments in watershed and coastal waters. OWOW facilitates states and tribes in the listing of impaired waters, which includes those that have been degraded by excess nutrients and are unable to meet the water quality standards set by states, territories, or authorized tribes. From this list, OWOW provides assistance for the development of total maximum daily loads (TMDLs), each of which is a calculation of the maximum amount of pollution that waters can receive and still meet water quality standards. Using TMDLs, OWOW is investigating the use of water quality trading programs as a possible way to control nutrient pollution. Additional relevant OWOW programs include the nonpoint source program, targeted watershed grants, the National Water Quality Inventory, and National

Aquatic Resource Surveys. OWOW also supports and provides leadership to the Mississippi Watershed/Gulf of Mexico Nutrient Task Force Coordinating Committee and participates and provides leadership to the Gulf of Mexico Alliance Water Quality and Nutrient Reduction teams. In addition, OWOW manages the Ocean Survey Vessel *Bold*, which supports coastal ocean assessment, monitoring and process-oriented research in US waters, including numerous systems impacted by nutrients and hypoxia.

EPA's National Estuary Program (NEP) was established by Congress in 1987 to improve the quality of estuaries of national importance. For information on the NEP, see National and Interagency Efforts, Section I.F.3.

Office of Science and Technology

The Office of Science and Technology (OST) produces regulations, guidelines, methods, standards, science-based criteria, and studies that are critical components of national programs that protect aquatic environments from the impacts of nutrient pollution. OST provides both technical and financial assistance to states to help them adopt numeric nutrient water quality standards. OST provides direct assistance to states by providing implementation guidance, addressing technical and policy issues that states raise, and providing technical information to support states rulemaking for standards. OST assists in building the scientific capacity for states through support for sampling and monitoring, training, data and statistical analysis, and modeling assistance for developing scientifically defensible nutrient criteria. OST also communicates the importance of managing nutrient pollution by integrating nutrient messages in communications and outreach products related to water quality standards.

I.D.3. Office of Research and Development

The mission of the Office of Research and Development is to conduct cutting-edge research and foster the sound use of science and technology to fulfill EPA's statutory and regulatory requirements. Its research laboratories are structured along the risk paradigm by focusing research on exposure, effects, risk assessment and risk management.

The National Exposure Research Laboratory

The National Exposure Research Laboratory (NERL) conducts research and development that leads to improved methods, measurements, and models to assess and predict exposures of humans and ecosystems to harmful pollutants and other conditions in air, water, soil, and food. NERL is conducting atmospheric deposition research and modeling of nitrogen, sulfur, and mercury as stressor inputs to water quality and ecosystem models to provide forecasts of changes in deposition due to management actions stemming from Clean Air Act regulations.

The National Health and Environment Effects Research Laboratory

The National Health and Environment Effects Research Laboratory (NHEERL) conducts research to determine the impacts of environmental stressors on human and ecosystem health as well as the degree to which those stressors cause harm. NHEERL conducts research that improves the science supporting nutrient criteria development in estuaries and coastal waters.

Case studies in Pensacola Bay, Florida, and Yaquina Bay, Oregon, have provided approaches that may be used to develop nutrient criteria and protect aquatic life. Additionally, NHEERL research on Gulf hypoxia is developing statistical, semi-empirical and coupled numerical model applications, data products, and other tools to quantify the biogeochemical processes and ecosystem responses to nutrient loads, quantify sources of uncertainty in nutrient load–hypoxia response relationships, and forecast the effects of nutrient management actions and load reduction targets on the extent of hypoxia.

In situ observations and analyses from coastal and continental shelf cruises provide the empirical bases for the numerical simulation models of eutrophication and hypoxia on the continental shelf.

Through an interagency agreement with the Naval Research Laboratory (NRL), a coupled water quality-sediment diagenesis model is being integrated into the NRL coastal ocean hydrodynamic model. The integrated numerical simulation model will provide improved mechanistic understanding of the processes regulating Gulf hypoxia.

The National Center for Environment Assessment

The National Center for Environment Assessment develops new methods for risk assessment and applies those methods to the cutting-edge issues that the US EPA faces.

The National Risk Management Research Laboratory

The National Risk Management Research Laboratory (NRMRL) conducts research and technical assistance to provide the scientific basis to support the development of strategies and technologies to protect and restore ground water, surface water, and ecosystems impacted by man-made and natural processes. NRMRL conducts nutrient fate and transport modeling, incorporating surface water-groundwater interactions, in the Mississippi River Basin. It also conducts research to improve manure management practices and reduce impacts of manure runoff on water quality. In addition, the Laboratory's research on best management practices focuses on evaluating how restoration techniques may enhance nutrient retention and attenuation in watersheds that are small enough for management practices to have a significant impact.

The National Center for Environmental Research

The National Center for Environmental Research (NCER) supports high-quality research that will provide the scientific bases for national environmental decisions. NCER supports cutting-edge, extramural research through competitions for grants, fellowships, and innovative small business research contracts. NCER's Science to Achieve Results (STAR) Program funds research grants and graduate fellowships related to various aspects of hypoxia, including nutrient source, fate, and transport. Overall, STAR awards about 150 research grants and 125 graduate fellowships each year. NCER also makes awards under joint Request for Applications with partnering agencies. These grants and fellowships have been awarded to universities and nonprofit research institutions in all 50 states, Guam, Puerto Rico, and the District of Columbia.

Environmental Monitoring and Assessment Program

The Environmental Monitoring and Assessment Program (EMAP) is a research program to develop the tools necessary to monitor and assess the status and trends of national ecological resources. EMAP's goal is to develop the scientific understanding for translating environmental monitoring data from multiple spatial and temporal scales into assessments of current ecological condition and forecasts of future risks to natural resources. The EMAP program also provides extramural funding through the STAR program. Through their Great Rivers Project, EMAP is providing an assessment of the ecological conditions in the Mississippi River Basin using a probability-based sampling design. This assessment will allow for state-level assessment of those rivers within or adjacent to a state's borders.

National Environmental Scientific Computing Center and the Environmental Modeling and Visualization Laboratory

The National Environmental Scientific Computing Center and the Environmental Modeling and Visualization Laboratory (EMVL) is an EPA-managed, contractor-operated facility designed to support EPA's scientific modeling and large-scale data management research programs through the application of high performance computing and visualization tools. The mission of EMVL falls in two categories: 1) to provide the high performance computing resources necessary to support environmental research of global proportions and improved science for the development of regulations and 2) to assist researchers in translating data into knowledge by using visual images to facilitate insight into that data. Currently, EMVL is supporting NHEERL research on Gulf hypoxia by assisting in the development of an integrated hydrodynamic, water quality, and sediment diagenesis simulation model for hypoxia in the Gulf of Mexico.

I.E. Department of Interior

I.E.1. US Geological Survey

Integrated Studies of Coastal Ecosystems

USGS conducts long-term assessments of the environmental and ecological conditions in aquatic ecosystems across the Nation in selected bays, estuaries, and other coastal environments. The research addresses a broad range of science issues related to environmental health and ecosystem management, including nutrient dynamics, algal blooms, and the role of terrestrial contributions of contaminants. The research in these ecosystems is coordinated closely with other Federal and local governmental agencies. Information about such programs are available at:

San Francisco Bay: http://sfbay.wr.usgs.gov/access/wqdata/
Chesapeake Bay: http://chesapeake.usgs.gov/
Gulf of Mexico: http://toxics.usgs.gov/hypoxia/
Tampa Bay: http://gulfsci.usgs.gov/tampabay/
Galveston Bay: http://gulfsci.usgs.gov/galveston/
Boston Harbor: http://woodshole.er.usgs.gov/project-pages/bostonharbor/
Puget Sound: http://puget.usgs.gov/

National Streamflow Information Program

The National Streamflow Information Program (http://water.usgs.gov /nsip) provides streamflow information at more than 7,000 stream gauges nationwide, funded by the USGS and over 800 local, state, and other Federal agencies, many of which represent downstream measuring sites for rivers discharging to the coast. These data are essential to understanding the role of and trends in freshwater contributions to coastal ecosystems and to quantifying the associated loadings of nutrients and other chemicals.

Water Quality Monitoring

The USGS National Stream Quality Accounting Network measures the annual transport of nutrients, dissolved solids, selected pesticides, and suspended-sediment from selected large rivers to coastal waters of the United States and from major inland sub-basins for priority large rivers, where the source origins of chemical loads are needed to guide management actions.

The network includes monitoring at approximately 18 major coastal rivers that contribute over 80 percent of the total discharge of streamflow, nitrogen, phosphorus, and suspended sediment to coastal waters of the United States.

Nineteen additional stations within the Mississippi River Basin are monitored to define the source origins of streamflow and nutrients ultimately delivered to the Gulf of Mexico to guide management actions to mitigate hypoxia in the northern Gulf of Mexico. Other major river systems where chemical loads are monitored at inland stream sites to guide management actions include the Chesapeake Bay and San Francisco Bay watersheds. USGS water quality monitoring of flow and chemical delivery to coastal waters is used to make estimates of national coastal delivery, such as those developed by the Heinz Center (http://www. heinzctr.org/ecosystems/2002report/national/ mvmt_n.shtml).

The National Water Quality Assessment (NAWQA) (http://water.usgs. gov/nawqa/) Program conducts assessments of the quality of the Nation's waters to provide an understanding of water quality conditions and how those conditions may vary locally, regionally, and nationally; whether conditions are getting better or worse over time; and how natural features and human activities affect those conditions.

NAWQA Program scientists collect and interpret data about surface- and groundwater chemistry, hydrology, land use, stream habitat, and aquatic life in parts or all of nearly all 50 states using a nationally consistent study design and uniform methods of sampling analysis. Water quality monitoring for national status and trends provide important data on the occurrence and distribution of nutrients and other water quality constituents in representative basins throughout the Nation. These data are used to develop the SPARROW and other water quality models that provide significant knowledge of the source areas of nutrients and the human activities that affect nutrient loads.

Groundwater Resources Program

Freshwater delivered through groundwater discharge to coastal waters is essential to ecosystem function and coastal circulation, and can have a significant influence on the health of these ecosystems.

The USGS Groundwater Resources Program (http://water.usgs.gov/ogw/ gwrp) conducts studies along the Atlantic coast related to ground-water discharge and its role in coastal salinity balance and the role of the associated nutrient contributions in coastal hypoxia.

Additional studies in these areas are conducted on a cooperative basis with local government agencies.

Nutrient Dynamics and Biochemical Cycling in Wetland and Riverine Ecosystems

USGS conducts research to improve the understanding of the biogeochemical cycling of nutrients in riverine and wetland ecosystems. Ecosystem management of ecosystems is emerging as an effective means of protecting natural resources and reducing nutrient inputs to receiving waters. Effective ecosystem management requires knowledge of the biogeochemical cycling of nutrients (requiring analysis of internal and external sources and sinks of nutrients) and understanding of the processes that control these sources and sinks.

In numerous aquatic ecosystems, many of these processes are poorly understood as are their potential interactions with other ecosystem properties, such as plant production. This research is being conducted primarily by the USGS National Research Program (http://water.usgs.gov/ nrp), the NAWQA Program, the National Wetlands Research Center (http://www.nwrc.usgs.gov), and the Upper Midwest Environmental Science Center (http://www.umesc. usgs.gov).

I.F. National and Regional Interagency Efforts

I.F.1. Integrated Ocean Observing Systems

The US Integrated Ocean Observing System (IOOS) (http://www. ocean.us) is a coordinated national network of observations and data transmission, data management, and communications intended to routinely and continuously acquire and disseminate quality controlled data and information on current and future states of the oceans and Great Lakes from the global scale of ocean basins to local scales of coastal ecosystems.

The IOOS is part of the US Integrated Earth Observing System, the US contribution to the Global Ocean Observing System (http://www.ioc-goos.org), and a contribution to the international Global Earth Observation System of Systems.

I.F.2. National Monitoring Network and the National Water Quality Monitoring Council

In response for calls for a National Water Quality Monitoring Network for streamflow and the associated chemical delivery to coastal waters, the National Water Quality Monitoring Council (http:// acwi.gov/monitoring), a subgroup of the Advisory Committee on Water Information, was charged with designing this network.

The design was developed by 80 representatives working through the National Water Quality Monitoring Council, including representatives from Federal, state, and local government organizations, universities, water associations, and the private sector. The Council is co-chaired by the USGS and the EPA, and its other Federal members include NOAA, Tennessee Valley Authority, USACE, USDA, and the remaining Department of the Interior (DOI) agencies. The Network will coordinate water quality monitoring across the Nation to provide a comprehensive database that supports assessment of the health of ocean, coastal, and Great Lakes resources. Additional information on the national monitoring network, including reports for the pilot phase conducted in three geographic areas (Delaware Bay, San Francisco Bay, and Lake Michigan) is available on at http://acwi.gov /monitoring/network.

I.F.3. National Estuary Program

EPA's National Estuary Program (NEP; http://www.epa.gov/owow /estuaries) is a national network of 28 programs working for collaborative solutions for estuaries designated by Congress as of critical importance. Created in 1987 under the Clean Water Act, the NEPs are charged with protecting and restoring US estuaries by engaging state and Federal agencies, nongovernmental organizations, and local communities in planning and management. EPA provides base funding and Federal oversight for each of the NEPs, but program management is typically the responsibility of an estuary stakeholder-based management committee.

Section 320 of the Clean Water Act (National Estuary Program) states that one of the main purposes of the NEP is to develop a comprehensive watershed ecosystem plan for the conservation and management of natural resources in NEP estuaries. NEPs are required to have inclusive stakeholder representation on any management or advisory committee. Representatives of state and Federal agencies (including EPA, USDA, and NOAA), interstate or regional agencies, local governments, industry and business, public and private educational institutions and the general public, should comprise such a committee.

I.F.4. Regional Efforts

Chesapeake Bay

The Chesapeake Bay Program (http://www.chesapeakebay.net) is a regional partnership charged with leading and directing restoration of the Chesapeake Bay. This partnership includes representatives from Maryland, Virginia, the District of Columbia, Pennsylvania, the Chesapeake Bay Commission, and citizen advisory groups.

The Federal government is formally represented by the US EPA, but a number of additional Federal partners are involved, including NOAA, USDA, National Park Service, USGS, USFWS, Federal Housing Administration, US Coast Guard, National Aeronautics and Space Administration (NASA), USPostal Service, General Services Administration, National Capital Planning Commission, Department of Defense, and Department of Education. Scientific guidance is provided by a Scientific Technical Advisory Committee.

The Chesapeake Bay Program is leading efforts to improve water quality in the Bay, especially related to nutrients, with the goal of improving dissolved oxygen levels and habitat quality through working with farmers, developers, homeowners and local governments to reduce pollutants. In 1998, the EPA and state members of the Bay Program (known as the Chesapeake Executive Council) signed the second Chesapeake Bay Agreement which called for a 40% reduction in nitrogen and phosphorus inputs by the year 2000. This nutrient reduction target was reaffirmed in 1992 and expanded to include Bay tributaries and a call for detailed tributary nutrient management strategies. A third agreement, *Chesapeake 2000*, also reaffirmed the 40% reduction target, but extended the target date to 2010 and outlined more specific strategies to meet the goal.

Northern Gulf of Mexico

The Mississippi River/Gulf of Mexico Watershed Nutrient Task Force was formed in 1997 to understand the cause and effects of eutrophication in the Gulf of Mexico and to coordinate and support nutrient management activities within the Mississippi River

Watershed. Members of the Task Force include executive level representatives from ten states, the EPA, NOAA, USACE, DOI, and USDA. Task Force members have designated senior managers within their respective agencies to serve on a Coordination Committee, which advises the Task Force on actions and manages workgroups and committees.

In 2001, the Task Force released an "Action Plan" (Mississippi River/Gulf of Mexico Watershed Nutrient Task Force 2001) to address Gulf hypoxia and nutrient pollution within the watershed, setting a goal of reducing the annual size of the hypoxic zone to 5,000 km^2 by 2015. Following a comprehensive scientific reassessment process that included an EPA Science Advisory Board Report (US EPA 2007), the Task Force released a "2008 Action Plan" (Mississippi River/Gulf of Mexico Watershed Nutrient Task Force 2008) that reaffirmed the goal of 5,000 km^2, although it was recognized the 2015 target may not be met, and placed additional focus on the management of phosphorus in addition to nitrogen. To reach this goal, the "2008 Action Plan" (Mississippi River/Gulf of Mexico Watershed Nutrient Task Force 2008) also recommended nitrogen and phosphorus reductions of at least 45%.

The Gulf of Mexico Alliance was formed in 2004 as a partnership between the five Gulf States (Florida, Alabama, Mississippi, Louisiana, and Texas) with the goal of increasing regional collaboration to protect and enhance the ecological and economic health of the Gulf. Although NOAA and EPA are designated as Federal coordinators, the Alliance actively partners with NSF, NASA, USACE, USDA, DOD, DOE, DOI, Department of Transportation, Department of Health and Human Services, and the Department of State. Following its formation, the Alliance released a three-year action plan in 2006 (http://www.gulfofmexicoalliance.org) that identified reducing nutrient inputs as one of five priority issue areas. Under the priority, increasing regional coordination for the development of TMDL's and addressing the hypoxic zone in the northern Gulf through an aligned Gulf States position were identified as key actions. Through its second action plan that began in 2009, the Alliance will continue to facilitate the development of coordinated nutrient criteria and reduce hypoxia in coastal waters and estuaries through increased coordination with Gulf Hypoxia Task Force and targeted watershed strategies.

Narragansett Bay

Significant cooperative efforts have been undertaken in Narragansett Bay to both understand and mitigate hypoxia. The Narragansett Bay Estuary Program (NBEP, http://www.nbep.org) was established in 1993 as part of the NEP. EPA, USDA, Rhode Island and Massachusetts state environmental agencies, academic partners (University of Rhode Island) and other nongovernmental stakeholders participate in the management of NBEP.

The NBEP also works closely with other Federal partners including NOAA, USACE, USGS, and USFWS. Since 1998, scientists from the NBEP, along with volunteer scientists from academic institutions and state agencies, have been conducting monthly (or more often) dissolved oxygen surveys. The data from these surveys are available at: http://www.geo.brown.edu/georesearch/ insomniacs/links.html.

A major menhaden fish kill in August 2003 was investigated by NBEP scientists in the context of the developing knowledge of dissolved oxygen dynamics in Narragansett Bay. The Rhode Island state legislature passed legislation the following year to reduce nitrogen inputs from wastewater treatment plants.

Great Lakes

In December 2004, the Great Lakes Regional Collaboration was launched as a result of a presidential executive order, creating a unique partnership of key members from Federal, state, and local governments, tribes, and other stakeholders for the purpose of developing a strategic action plan. The Federal members of the Task force are cabinet level secretaries of the Departments of State, Commerce, Defense, Agriculture, Housing and Urban Development, Transportation, and the Interior; EPA; and the Council on Environmental Quality. This strategic plan, the *Great Lakes Regional Collaboration Strategy*, was signed in December 2005 and is intended to build upon the extensive regional efforts to date, working together toward a common goal of restoring and protecting the Great Lakes ecosystem for this and future generations. Although hypoxia is not particularly emphasized, nutrient reduction is an important part of the strategy. http://www.glrc.us.

Puget Sound

A dramatic 2003 fish kill in Hood Canal which was caused by the temporary upwelling of hypoxic waters precipitated strong public and scientific interest in understanding hypoxia dynamics in the Puget Sound ecosystem.

The goal of the Hood Canal Dissolved Oxygen Program (HCDOP) supported through Congressional funding to the University of Washington is to determine the sources of low dissolved oxygen in Hood Canal and its effect on marine life. HCDOP has worked with local, state, Federal, and tribal government policy makers to develop potential corrective actions that will restore and maintain a level of dissolved oxygen that will not stress marine life. HCDOP is a partnership of 28 organizations that conducts monitoring and analysis and develops potential corrective actions to address the low dissolved oxygen problem in Hood Canal, including Federal (NOAA, USGS, USFWS. USACE, EPA and US Navy), state, and tribal entities. http://www.hoodcanal. washington.edu. In 2007, the state legislature created the Puget Sound Partnership which took over the role of the Puget Sound Action Team.

The goals of the partnership are to prioritize cleanup and improvement projects; coordinate Federal, state, local, tribal, and private resources; encourage cooperation; and make decisions based on sound science—some of which will be provided by the HCDOP. Several Federal agencies are represented on the partnership councils, including NOAA, Navy, USGS, USFWS, and EPA. http://www. psp.wa.gov.

Long Island Sound

The Long Island Sound Study (LISS) was formed by EPA, the State of New York, and the State of Connecticut to cooperatively address the degraded ecosystem within Long Island Sound. This partnership also involves relevant nonprofit organizations, concerned citizens, as well as NOAA, EPA, USGS, USFWS, and USDA. The LISS completed a Comprehensive Conservation and Management Plan that identified hypoxia in the Sound as one of seven priority issues. To address hypoxia, in 1998 the LISS adopted a 58.5% nutrient reduction target for nitrogen by 2014, and nutrient management plans for New York and Connecticut were formally approved by EPA in 2001.

These nutrient management plans are focusing on both nonpoint and point source pollution. Scientific support for the management plan is developed through the LISS Scientific Technical Advisory Committee, with financial support provided through a cooperative agreement among the EPA and the New York and Connecticut Sea Grant Programs. The LISS also funds the monitoring of hypoxia, water quality, and plankton. http://www. longislandsoundstudy.net.

APPENDIX II. GEOGRAPHIC CASE STUDIES

Introduction and Background to Case Studies

The case studies presented here highlight selected coastal ecosystems that are geographically diverse and represent the spectrum of estuarine and coastal ocean ecosystems affected by hypoxia. The selected cases describe the range of circumstances, understanding, and scientific and management approaches across the United States (Figure A1). They provide examples of monitoring and research programs that have been or are being used to develop management plans.

In some instances the outcomes of implemented management measures are also presented and should serve as encouragement that nutrient-related dissolved oxygen problems can be improved with proper management. The exception is hypoxia along the Oregon coast, which has occurred since 2002 and is more related to climate impacts on upwelling-favorable winds on the west coast. Yet, coastal management is still pertinent to the Oregon shelf which has thriving coastal fisheries negatively impacted by hypoxia. These case studies provide the basis for successful management approaches that may be used in other systems in the United States.

LONG ISLAND SOUND

Physical Description of the System

Long Island Sound is a large (3,056 km^2) glacial outwash estuary, shared by the States of Connecticut and New York (Figure 1). Its unique configuration connects it with the Atlantic Ocean via the Race in the east and via the East River in the west.

The Connecticut River, very near the Sound's eastern terminus, contributes about two thirds of its freshwater input. The Housatonic and Thames Rivers also contribute significant volumes of fresh water to this system.

The influence of the East River (a tidal strait) promotes a stratified salinity structure in the western sound especially in the spring runoff period when freshwater from the Hudson River basin is transported into western Long Island Sound. Salinity variability is less distinct in the eastern sound where salinities tend to be higher.

Tidal amplitude ranges from about two meters in the west to less than one meter in the east. The Sound is moderately flushed, with mean residence times of two to three months. A highly developed watershed contributes to low dissolved oxygen problems (from Bricker et al. 1997, 2007).

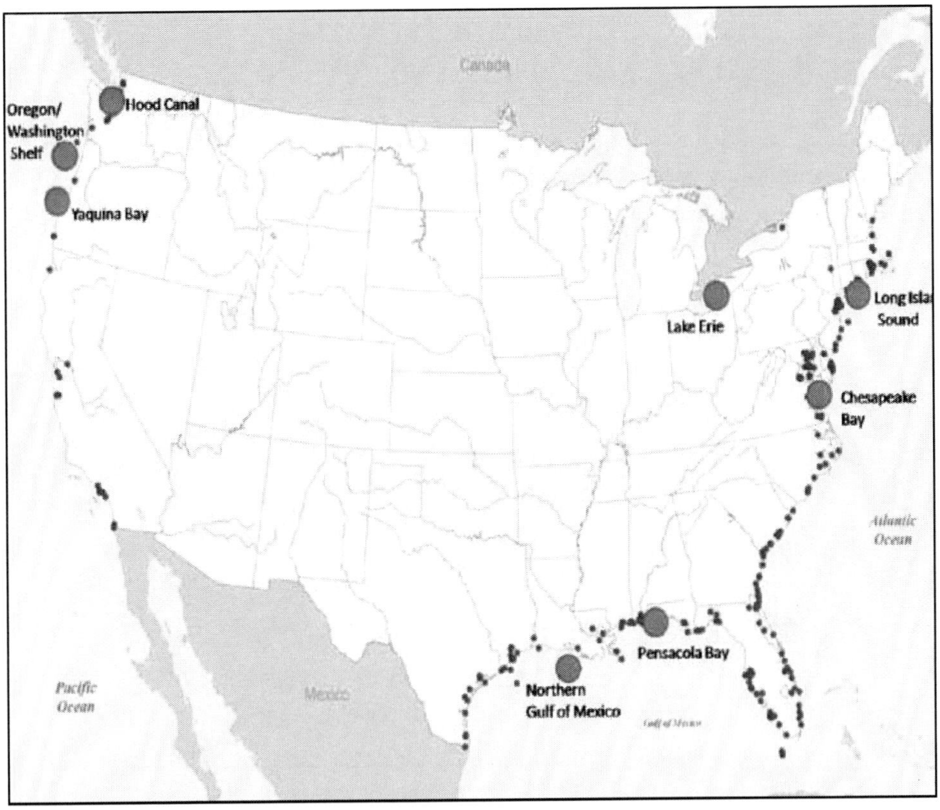

Figure A1. Geographic locations of hypoxia case studies. Large blue dots represent cases studies; red dots represent systems that have a documented hypoxia issue.

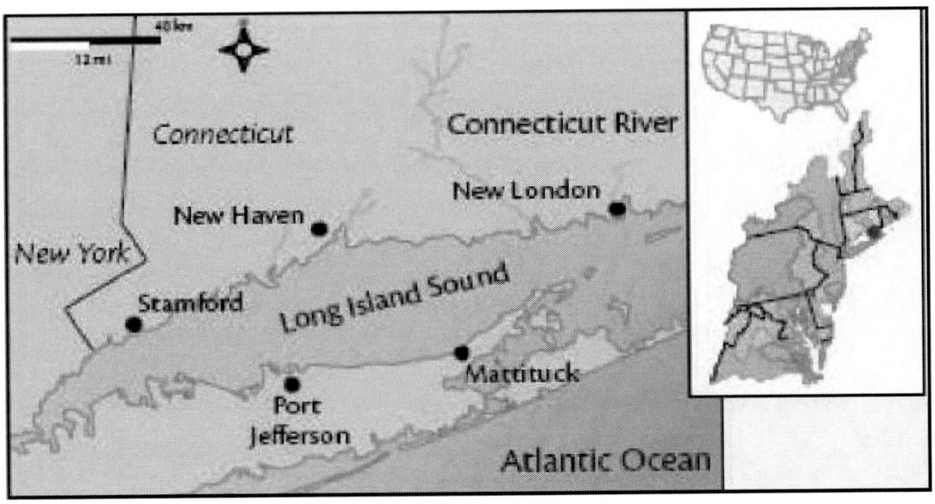

Figure 1. Location of Long Island Sound (P. Stacey, CT DEP).

History of Hypoxia (Issue, Causes, Economic, and Ecosystem Impacts)

Long Island Sound has a large and highly developed watershed (Table A1). Nitrogen contributions from the watershed, combined with strong summer thermal stratification in its western half, renders Long Island Sound susceptible to seasonal hypoxia.

Since 1985, the causes and effects of hypoxia have been the subject of intensive monitoring, modeling, and research through the Long Island Sound Study (LISS), part of the EPA National Estuary Program (see Appendix I, Interagency Efforts).

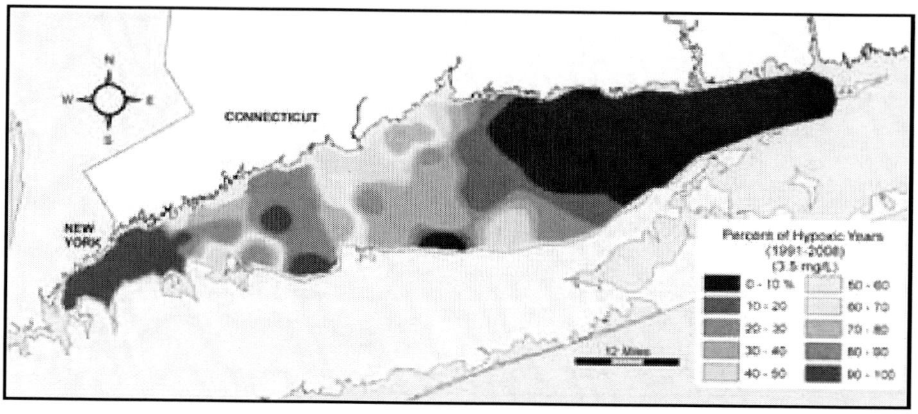

Figure 2. The frequency and location of hypoxia in Long Island Sound bottom waters from 1991-2008 (P. Stacey CT DEP).

Hypoxia most seriously affects the strongly stratified western half of the Sound where dissolved oxygen concentrations fall well below Connecticut and New York's water quality standards each summer (Figure 2). Dissolved oxygen levels below 3 mg/L are usually observed, levels below 2 mg/L are not uncommon, and during some years portions of the Sound's bottom waters become anoxic (<1 mg/L; Figures 2 and 3).

Long Island Sound is surrounded by a highly urbanized landscape, including New York City in the west and large, sprawling populated areas in Long Island and Connecticut. Primary sources of nitrogen include sewage treatment plants, nonpoint source runoff, and atmospheric deposition, all driven by human influences in the watershed and airshed (Figure 4). Monitoring of Long Island Sound conducted by the Connecticut Department of Environmental Protection on behalf of the LISS has shown an annual recurrence and persistence of hypoxia over the last 15 years. Despite significant reductions in nitrogen loads by both Connecticut and New York under a total maximum daily load (TMDL) approved in 2001, dissolved oxygen improvements have been slow and masked by weather-driven variability and effects of climate change.

Considering that nature contributes, at most, about 10,000 metric tons of nitrogen every year to the Sound from weathering and nitrogen fixation in the watershed, the nearly 38,000 metric tons per year added by more than 100 sewage treatment plants located along the coast and throughout the drainage basin have greatly enriched the ecosystem (Figure 5).

Table A1. Comparison of Physical Systems Represented by Case Studies in Appendix II

Site	Waterbody area (km2)	Watershed area (km2)	Average depth (m)	Tidal height (m)	Freshwater inflowm3/d	Average salinity (psu)	Watershed population	NEEA Results*: Dissolved oxygen rating	
								Early 1990s	Early 2000s
Long Island Sound	3,259	12,773	20	1.9	1.55×10^7	28	4.91×10^6	Moderate	High
Lake Erie	25,744	--	19	0	---	0	---	---	---
Chesapeake Bay Mainstem	6,974	79584	7.3	0.45	1.05 x 108	16	6.41×10^6	High	High
Pensacola Bay	477	17650	3.0	0.42	2.58 x 107	18	3.71×10^5	Moderate	Low
Northern Gulf of Mexico	31,743	2,968,304	20	0.29	1.53 x 109	29	7.30×10^7	High	High
Yaquina Bay	14	634	2.1	1.9	6.60 x 105	22	6.03x103	Low	Low
Oregon Shelf	26,600	---	---	3-4 (nearshore)	5.6 x 108	32	---	---	---
Hood Canal	396	2768	70	2.2	7.45 x 106	26	3.49×10^4	Moderate	High

* From Bricker, S., B. Longstaff, W. Dennison, A. Jones, K. Boicourt, C. Wicks and J. Woerner. 2007. Effects of Nutrient Enrichment in the Nation's Estuaries: A Decade of Change, National Estuarine Eutrophication Assessment Update. NOAA Coastal Ocean Program Decision Analysis Series No. 26. National Centers for Coastal Ocean Science, Silver Spring, MD. 322 pp. http://ccma.nos.noaa.gov/news/feature/Eutroupdate.html.

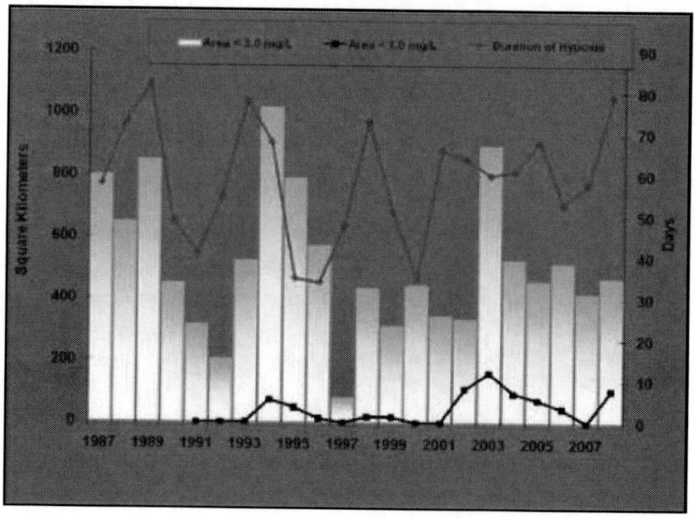

Figure 3. Areal extent and duration of Long Island Sound hypoxia, 1987-2007. The average areal extent is 521 km2 and average duration is 58 days. (P. Stacey CT DEP).

Another 12,500 metric tons of nitrogen are contributed each year from nonpoint sources coming from excessive fertilizer added to lawns or agricultural crops, atmospheric emissions (from automobiles, power plants, and industry), and human and animal wastes (including home septic systems). Population continues to grow within the already densely populated Long Island Sound basin. This growth contributes to the nitrogen load through sewage treatment plants as well as through various nonpoint sources, and the impact is large. Since 1985, Connecticut's land conversion rate to developed uses was 11.3% for a population that had grown about 8.6%. Per capita consumption of land is outstripping population growth in Connecticut and throughout the basin.Considering that nature contributes, at most, about 10,000 metric tons of nitrogen every year to the Sound from weathering and nitrogen fixation in the watershed, the nearly 38,000 metric tons per year added by more than 100 sewage treatment plants located along the coast and throughout the drainage basin have greatly enriched the ecosystem (Figure 5). Another 12,500 metric tons of nitrogen are contributed each year from nonpoint sources coming from excessive fertilizer added to lawns or agricultural crops, atmospheric emissions (from automobiles, power plants, and industry), and human and animal wastes (including home septic systems). Population continues to grow within the already densely populated Long Island Sound basin. This growth contributes to the nitrogen load through sewage treatment plants as well as through various nonpoint sources, and the impact is large. Since 1985, Connecticut's land conversion rate to developed uses was 11.3% for a population that had grown about 8.6%. Per capita consumption of land is outstripping population growth in Connecticut and throughout the basin.

Nitrogen enrichment, coupled with the Sound's sensitivity to hypoxia due to relatively long residence time and seasonally strong stratification, leads to unhealthy conditions that are environmentally and economically costly for the Sound and its users. Furthermore, submerged aquatic vegetation (SAV) decline has been observed in many eastern Long Island Sound embayments and is believed to be linked to nitrogen enrichment. Nitrogen reductions to protect SAV will be much more stringent and difficult to attain than for low dissolved oxygen (Stacey 2007).

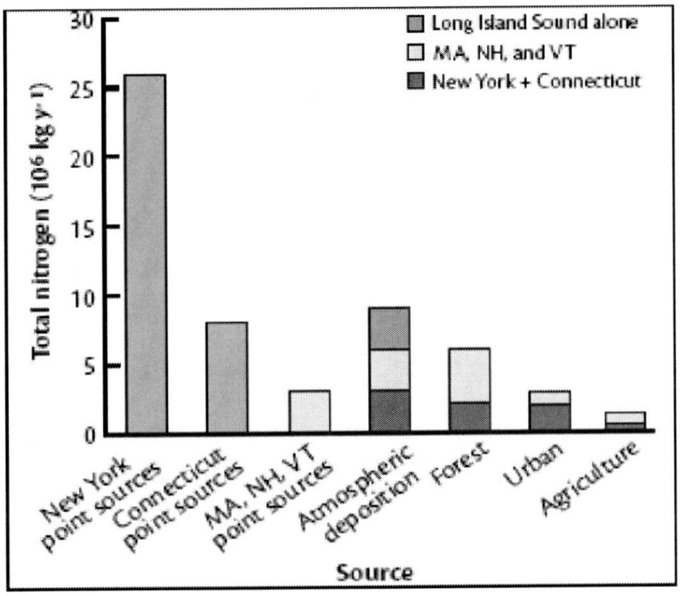

Figure 4. Sources of nitrogen to Long Island Sound – baseline condition (P.Stacey, CT DEP).

SCIENCE AND MANAGEMENT ACTIONS (TO DATE AND PLANNED)

Since 1985, the causes and effects of hypoxia have been the subject of intensive monitoring, modeling, and research through the LISS. Through the Long Island Sound partnership, a dissolved oxygen TMDL was completed by Connecticut and New York and approved by US EPA in 2001. Both states have aggressively pursued sewage treatment plant nitrogen control using biological processes, and nitrogen loads are trending downward. New York has relied upon traditional permitting programs to limit individual sewage treatment plants while Connecticut has instituted a statewide nitrogen trading program, called Nitrogen Credit Exchange, for 79 municipal sewage treatment plants (CT DEP 2007). The trading program works as an economic engine, forcing action towards the most cost-effective and environmentally beneficial projects. Collectively, the two states have accomplished a 30% reduction in sewage treatment plant nitrogen loads towards the TMDL target of 60–65% by 2014. The nitrogen trading program has proven to be a viable alternative for improving dissolved oxygen conditions in Long Island Sound and has accelerated the schedule to meet the TMDL wasteload allocation deadline. Potential savings with nitrogen trading are in the \$200-400 million range over individual permitting approaches. The Nitrogen Credit Exchange could serve as a working model for incentive-based trading in other states.

Connecticut and New York are also relying on stormwater permitting and nonpoint source programs to meet a 10% reduction target for urban and agricultural lands. However, the nonpoint sources are more difficult and costly to control, especially atmospheric deposition, much of which originates from jurisdictions outside of Connecticut and New York. If attained, promised reductions from Federal Clean Air Act initiatives will help Long Island Sound tremendously.

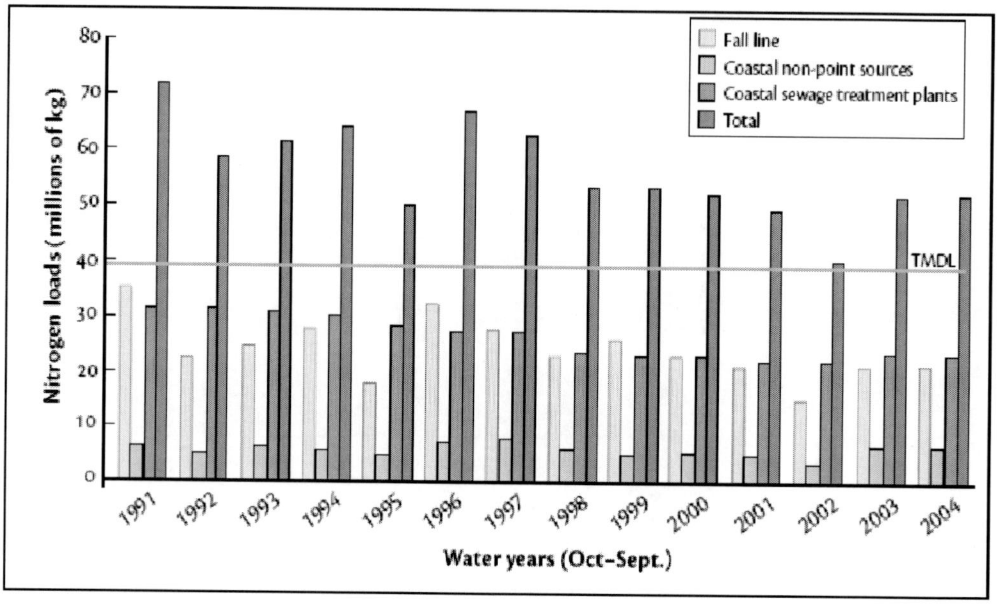

Figure 5. Estimated nitrogen loads to Long Island Sound, 1991-2004 (P. Stacey, CT DEP).

The LISS is also working with Massachusetts, New Hampshire, and Vermont, states that share the Long Island Sound watershed, to ascertain what level of reduction might be achieved in a cost-effective manner by those states.

FUTURE OUTLOOK

Population continues to grow within the already densely populated Long Island Sound basin. This growth contributes to the nitrogen load from sewage treatment plants as well as nonpoint sources. Despite the high land conversion rate since 1985, nitrogen loads have been reduced by some degree, largely through active management of sewage treatment. However, nutrient-related water quality problems have not improved as much as anticipated, in part, because of the high rate of per capita land consumption (sprawl) relative to population growth. As a result, management measures are offset by the increase in nutrient inputs related to population increases. In addition, the effects of climate change appear to have strengthened stratification in the Sound, exacerbating hypoxia.

REFERENCES

Bricker S, Clement C, Frew S, Harmon M, Harris M, Pirhalla D. 1997. NOAA's Estuarine Eutrophication Survey.Volume 2: Mid-Atlantic Region. Silver Spring, MD. Office of Ocean Resources Conservation Assessment. 51 pp.

Bricker S, Longstaff B, Dennison W, Jones A, Boicourt K, Wicks C, Woerner J. 2007. Effects of Nutrient Enrichment in the Nation's Estuaries: A Decade of Change, National Estuarine Eutrophication Assessment Update. NOAA Coastal Ocean Program Decision

Analysis Series No. 26. National Centers for Coastal Ocean Science, Silver Spring, MD. 322 pp. http://ccma.nos.noaa.gov/news/feature/Eutroupdate.html.
CT DEP. 2007. Connecticut's Nitrogen Credit Exchange – An Incentive- based Water Quality Trading Program. The Connecticut Department of Environmental Protection, Bureau of Waer Protection and Land Reuse. Hartford, CT.
Stacey PE. Personal Communication. Connecticut Department of Environmental Protection.
Stacey PE. 2007. Long Island Sound, Ct & NY: Point source reductions lessened hypoxia in 1990s. In: Bricker S, Longstaff B, Dennison W, Jones A, Boicourt K, Wicks C, Woerner J. 2007. Effects of Nutrient Enrichment in the Nation's Estuaries: A Decade of Change, National Estuarine Eutrophication Assessment Update. NOAA Coastal Ocean Program Decision Analysis Series No. 26. National Centers for Coastal Ocean Science, Silver Spring, MD. 322 pp. http://ccma.nos.noaa.gov/news/ feature/ Eutroupdate.html.

Lake Erie

Physical Description of the System

Lake Erie is the shallowest of the Great Lakes with an average depth of 19 meters. It has three identifiable sub-basins: the eastern, central, and western basins. Lake Erie is deepest in the eastern basin and becomes increasingly shallow towards the west. Because it is the shallowest of the Great Lakes, it warms more quickly in the summer and cools more rapidly in the fall. The shallow depth and warm temperatures lead to high productivity. Thermal stratification, caused by warming of the upper layer, occurs throughout the Lake every summer. However, vulnerability to hypoxia is not consistent basin to basin. Eighty percent of the Lake's water inflow comes from the Detroit River, but most of the nutrient and sediment loading comes from other tributaries, notably the Maumee River. Lake Erie receives the highest sediment loads of all the Great Lakes (Letterhos 2007).

History of Hypoxia (issue, causes, economic and ecosystem impacts)

The central basin of Lake Erie is the most vulnerable to hypoxia because it is sufficiently deep to develop thermal stratification, but the volume of water below the thermocline (or temperature gradient) is small enough that oxygen is depleted quickly. The water below the thermocline is cut off from oxygen in the surface water due to the stratification. The bottom waters in the central basin frequently go hypoxic during late summer through early fall (Charlton 1980, Rosa and Burns 1987). In contrast, the much deeper eastern basin, which also develops thermal stratification, has a much higher water volume below the thermocline, so the oxygen is not depleted before the water column mixes when the cold weather returns in the fall. The shallower western basin experiences stronger wind-driven circulation that prevents stratification, and thus hypoxia, during most summers.

Although late summer hypoxia is a natural phenomena in Lake Erie probably dating back thousands of years (Delorme 1982), evidence suggests that the summer oxygen depletion rates increased during the 1950s and 1960s due to increased phosphorus loadings from point and nonpoint sources. Even the shallower western basin developed hypoxia during low wind periods (Hartman 1972, Leach and Nepszy 1976, Beeton 1963). Concerted phosphorus reduction programs led to improved conditions until late 1990s. However, the extent of bottom water hypoxia has returned to pre-action levels. The cause is much less clear, but likely related to increases in nutrient inputs from nonpoint sources (Richards, Unpublished Data) (Figure 1). Total phosphorous and soluble reactive phosphorous concentrations in the mainstem of the Lake have remained low, but levels in the nearshore areas are rising. Phosphorus, however, may no longer be the limiting nutrient; nitrogen and carbon may be playing greater roles (Letterhos 2007).

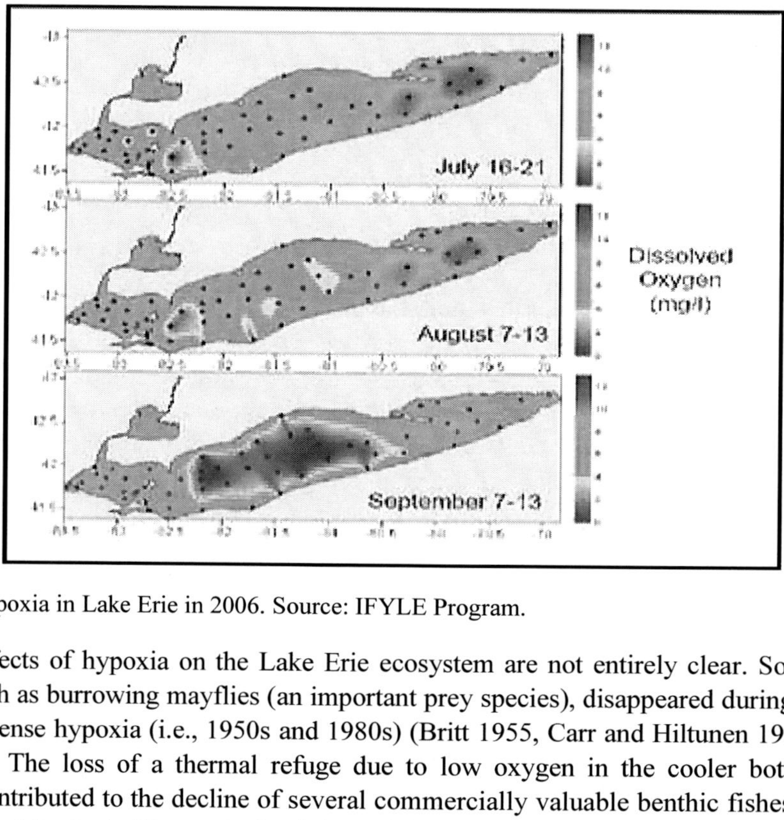

Figure 1. Hypoxia in Lake Erie in 2006. Source: IFYLE Program.

The effects of hypoxia on the Lake Erie ecosystem are not entirely clear. Some benthic species, such as burrowing mayflies (an important prey species), disappeared during periods of the most intense hypoxia (i.e., 1950s and 1980s) (Britt 1955, Carr and Hiltunen 1965, Krieger et al.1996). The loss of a thermal refuge due to low oxygen in the cooler bottom waters probably contributed to the decline of several commercially valuable benthic fishes (e.g., lake whitefish and burbot). The arrival of the zebra mussel in the 1980s caused tremendous ecological change and may be contributing to hypoxia. The effect on Lake Erie fisheries is currently an area of active research (see Appendix I, NOAA/GLERL International Field Years on Lake Erie).

Science and Management Actions (to Date and Planned)

The Great Lakes Water Quality Agreement of 1978 (http://www.epa.gov /glnpo/glwqa/ 1978/index.html) was a binational agreement between Canada and the United States which set strict limits on phosphorus loadings, a goal which was met in the 1980s (Hawley et al. 2006). The agreement targeted inputs from wastewater treatment plants, limited the use of phosphorus in detergents, and led to development of best management practices to reduce phosphorus from agricultural runoff. By the late 1980s, the goals outlined in the agreement were largely achieved and the incidence and extent of hypoxia significantly decreased (Hawley et al. 2006).

However, the severity of hypoxia and levels of phosphorus in the lake began to increase in the 1990's, although the cause is relatively unknown (Hawley et al. 2006).

In 1987 the governments of Canada and the United States also made a commitment, as part of the Great Lakes Water Quality Agreement, to develop a Lakewide Management Plan for the Great Lakes. The Lakewide Management Plan (LaMP) for Lake Erie (http:// www.epa.gov /greatlakes /erie.html) is coordinated by Federal, state, and provincial government agencies in the two countries. Under the guidance of these agencies, the LaMP unites a network of stakeholders in actions to restore and protect the Lake Erie ecosystem.

The LaMP provides an opportunity to link efforts to work toward the common goal of restoring Lake Erie for future generations.

In 2005, the International Field Years on Lake Erie (IFYLE) was initiated by researchers from the United States and Canada to conduct comprehensive research in the Lake, with an initial focus on hypoxia and harmful algal blooms. The primary objectives of this effort focus on quantifying the spatial extent of hypoxia to enable the development of forecasts as well as to determine the ecological consequences of hypoxia to Lake Erie living resources.

In December 2005, the Great Lakes Regional Collaboration (GLRC) released a Strategy to Restore and Protect the Great Lakes (http://www.glrc. us/strategy.html), which addressed nonpoint source pollution in the recommendations. More recently, the President's 2010 Budget provided $475 million in EPA's budget for a new Environmental Protection Agency-led, interagency Great Lakes Restoration Initiative. This Great Lakes Restoration Initiative is guided by the GLRC strategy and will target the most significant problems in the region, including invasive aquatic species, nonpoint source pollution, and contaminated sediment.

Future Outlook

The IFYLE will hopefully elucidate the most important factors contributing to the increase in hypoxia since the mid-1990s. Further, the impressive international cooperative effort in the region, represented by the Great Lakes Water Quality Agreement, LaMPs, Great Lakes Regional Collaboration (http://www. glrc.us/), and the Great Lakes Restoration Initiative, to identify and reduce stressors in Lake Erie and the other Great Lakes improves the chance of reducing hypoxia in the future. Management actions that reduce nutrient loadings in Lake Erie may have positive ramifications for both freshwater harmful algal blooms and hypoxia.

REFERENCES

Beeton AM. 1963. Limnological survey of Lake Erie in 1959 and 1960. Great Lakes Fishery Commission Technical Report 6: 1-32.

Britt NW. 1955. Stratification in western Lake Erie in summer of 1953: effects on the *Hexagenia* (Ephemeroptera) population. *Ecology* 36: 239-244.

Carr JF, Hiltunen K. 1965. Changes in the bottom fauna of western Lake Erie from 1930 to 1961. *Limnology and Oceanography* 10: 551-569.

Charlton MN. 1980. Oxygen depletion in Lake Erie: has there been any change? *Canadian Journal of Fisheries and Aquatic Sciences* 37: 72-81.

Delorme LD. 1982. Lake Erie oxygen: the prehistoric record. Canadian *Journal of Fisheries and Aquatic Sciences* 39: 1021-1029.

Hartman WL. 1972. Lake Erie: effects of exploitation, environmental changes and new species on the fishery resources. *Journal of the Fisheries Research Board of Canada* 29: 899-912.

Hawley N, Johengen TH, Rao YR, Ruberg SA, Beletsky D, Ludsin SA, Eadie BJ, Schwab DJ, Croley TE, Brandt SB. 2006. Lake Erie Hypoxia Prompts Canada-US Study. *EOS Transactions* 87(32): 313-324.

Krieger KA, Schloesser DW, Manny BA, Trisler CE, Heady SE, Ciborowski JJH, Muth KM. 1996. Recovery of burrowing mayflies (Ephemeroptera: Ephemeridae: *Hexagenia*) in western Lake Erie. *Journal of Great Lakes Research* 22: 254-263.

Leach JH, Nepszy SJ. 1976. The fish community in Lake Erie. *Journal of the Fisheries Research Board of Canada* 33: 622-638.

Letterhos J. 2007. Lake Erie Background from a Nutrient Perspective Prepared for the Ohio Phosphorus Task Force. Ohio EPA http://www.epa.state. oh.us/dsw/lakeerie/ptaskforce/ LakeErieBackground.pdf

Richards RP. Unpublished Data. Heidelberg University, Tiffin, Ohio.

Rosa F, Burns NM. 1987. Lake Erie central basin oxygen depletion changes from 1929-1980. *Journal of Great Lakes Research* 13: 684-696.

CHESAPEAKE BAY

Physical Description of the System

The Chesapeake Bay is a drowned river valley that is almost 300 km long with a relatively deep (20- 30 m) and narrow (1-4 km) central channel (Kemp et al. 2005) (Figure 1). Broad shallow areas flank this central channel over its entire length (Boicourt et al. 1999), and the mean depth of the estuary is 6.5 m (Kemp et al. 2005). Typically, the Bay is stratified with the surface freshwater isolating the deep and saline channel waters and limiting vertical mixing. Although occasionally strong wind events may cause some vertical exchange, stratification quickly returns due to the north-south salinity gradient (Boicourt 1992). There is a natural tendency for oxygen depletion in the deeper waters of the Bay due to the narrow and deep channel, persistent stratification, wide and shallow (and productive) sills flanking the channel, and a long water residence time (Boicourt 1992).

Relative to other estuaries, the Chesapeake Bay watershed is large (172,000 km^2) compared to the size and volume of the estuary (Bricker et al. 1999). The shoreline is long (18,800 km) and many rivers drain the watershed into the Bay. The largest tributary is the Susquehanna River which accounts for 41% of the watershed. The Bay is closely connected to its watershed which delivers freshwater to the Bay at an average rate of 2,300 m^3/s. Year-to-year fluctuations in river flow result in highly variable inputs of freshwater, nutrients, and sediment. The volume and nutrient composition of these flows influence stratification and productivity in the Bay (Kemp et al. 2005). This watershed is heavily populated, with an average density of 156 people per km2. In addition to the major population centers surrounding and connecting Baltimore and Annapolis, Maryland; Washington D.C.; and Richmond, Virginia; agriculture is a prominent feature of the watershed's landscape (Bricker et al. 2007).

History of Hypoxia (Issue, Causes, Economic and Ecosystem Impacts)

A recent assessment of eutrophic conditions in the Bay showed that it is among the most impacted estuaries in the United States. Nitrogen loads are high, and the Bay is susceptible to development of nutrient-related problems (Bricker et al. 2007).

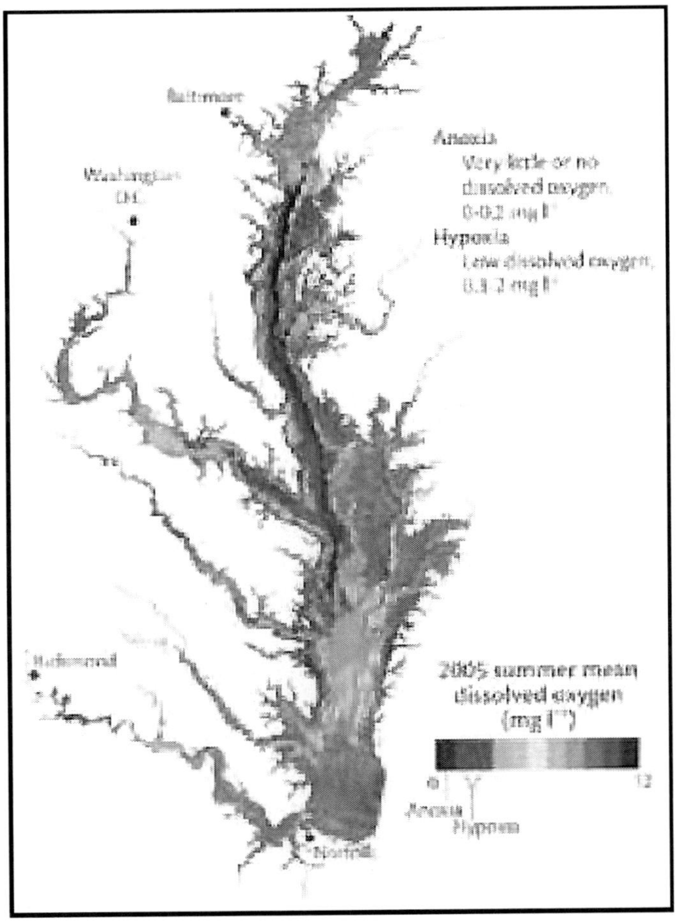

Figure 1. Dissolved oxygen conditions in the Chesapeake Bay in 2005 http://www.chesapeakebay.net/bayforecastspring2006.htm www.eco-check.org.

Eutrophic symptoms occur periodically or persistently and over extensive areas. These symptoms include high concentrations of chlorophyll-*a*, hypoxia, loss of submerged aquatic vegetation, nuisance and toxic algal blooms, and excessive growth of macroalgae. Throughout the Bay and its tributaries, symptoms have either worsened or not changed between 1999 and present (Bricker et al. 2007) (Figure 2).

Primary sources of nutrients include agriculture, wastewater, and urban runoff. Other sources include septic tanks, sewer overflows, and atmospheric deposition. The sources of nutrient inputs and their transport have been described in detail by Boynton et al. (1995) who determined that nitrogen and phosphorus inputs from the watershed and atmosphere increased by six to eight fold and 13 to 24 fold, respectively, between pre-colonial times and the mid-1980s. About one- fourth of the nitrogen and one-third of the phosphorus are from point sources, with the rest from nonpoint terrestrial and atmospheric sources (see Chapter 1, Figure 3).

Several seasonally varying factors influence the dynamics of oxygen depletion from bottom-waters of the Bay. The decline in dissolved oxygen in the spring appears to be controlled primarily by physical processes (e.g., stratification). Whereas, the late spring decline and the extent of summer hypoxia appear to be influenced by the level of phytoplankton production and by water temperature (Hagy et al. 2004), both of which influence the respiration rate.

On an annual basis, the amount of freshwater inflow is a good predictor of bottom-water oxygen depletion, because it controls water column stratification and, therefore, the rate of oxygen replenishment through vertical mixing. Freshwater runoff also delivers nutrients, which fertilize phytoplankton growth. Year-to-year variations in runoff have also been correlated with production and sedimentation of organic matter (Boynton and Kemp 2000) and increased dissolved oxygen demand.

Evidence from sediment studies in the Chesapeake Bay suggests that hypoxia has increased in occurrence over many decades (Kemp et al. 2005). Geochemical and paleontological methods were used to interpret this indirect ecological history (Brush 1984), and results suggest that eutrophication increased starting in the 17th century, coinciding with settlement by Europeans.

Eutrophication accelerated during the 20th century as indicated by a pronounced shift in the ratio of planktonic to, typically, benthic diatoms, reflecting a decline in both water clarity and benthic algal production (Cooper and Brush 1993). Increases in fossil abundance of hypoxia-tolerant species (Karlsen et al. 2000), iron pyrite (Cooper and Brush 1991, 1993), and geochemical changes (Zimmerman and Canuel 2002) provide additional sedimentary evidence for development of hypoxia in the 20th century.

Year-to-year variations in hypoxia extent and intensity are correlated with the amount of nitrogen entering the Bay. The amount of hypoxia generated from a given level of nitrogen loading has, however, been more severe in recent years relative to the past (Hagy et al. 2004). Since the early 1990s, the volume of hypoxic water has increased, while nitrogen loading has leveled or declined (Boesch et al. 2001, Hagy et al. 2004, Langland et al. 2006)).

In the mid-1980s, there was a shift in the relationship between hypoxia and nitrogen loading, suggesting that the Bay has passed an ecological threshold and is now more susceptible to eutrophication processes. This shift may have reduced the capacity of natural processes to both oxygenate the water column and trap nutrients in the sediments (Kemp et al. 2005).

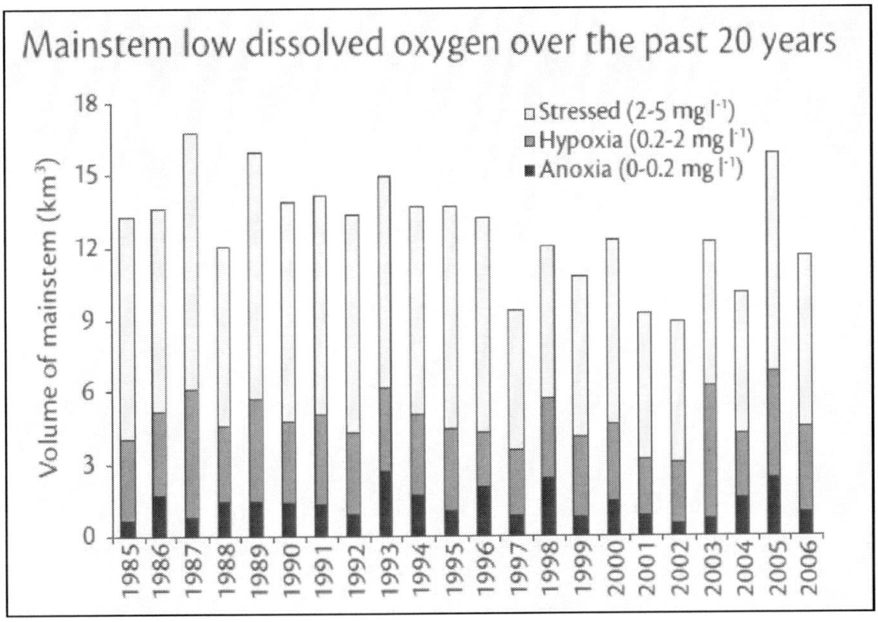

Figure 2. Volume of Chesapeake Bay affected by hypoxia/anoxia by year. http://www.chesapeakebay.net/bayforecastspring2006.htm www.eco-check.org.

Further compounding the effects of eutrophication has been the dramatic loss of oyster beds (currently biomass is approximately 1% of the 19th century levels, Newell 1988). Oysters strongly influence the Bay ecosystems through an ability to assimilate nutrients and rebound from hypoxic events (Kemp et al. 2005).

Oysters reduce concentrations of phytoplankton and other suspended particles through filter-feeding, thereby increasing light levels reaching the sediment (e.g., Cohen et al. 1984, Newell and Koch 2004). The loss of oysters has reduced water clarity, which in turn has negatively impacted the growth of vital submerged aquatic vegetation. It has been estimated that at the height of ecosystem health, Bay waters were filtered once every four days, but the depleted filter feeder populations now require more than a year to filter the same volume (Newell 1988).

Although heavy fishing pressure and other stressors in Chesapeake Bay make it difficult to isolate effects of eutrophication on fish, an increased ratio of pelagic (water column-dwelling) to demersal (bottom-dwelling) fish species documented for the Chesapeake Bay (Kemp et al. 2005) is indicative of increased eutrophication. Some of the most noticeable changes were the increase in pelagic Atlantic Menhaden, and the decline in demersal blue crab and oyster landings.

This shift from demersaldominated to pelagic-dominated fisheries has been observed for other coastal systems and is attributed, in part, to bottom-water hypoxia (e.g., Caddy 1993). It is not clear how hypoxia influences the habitat requirements of particular fish and invertebrates, but some species, such as sturgeons, can no longer reproduce or rear young in the Bay due to the lack of habitat with adequate oxygen and temperature levels (Niklitschek 2001, as cited in Kemp et al. 2005).

Whereas impacts to fisheries have been difficult to determine, impacts of hypoxia on benthic animals have been well documented. The diversity, abundance, and productivity of many benthic animals is affected by seasonal hypoxia in the Bay, particularly in deeper water (Holland et al. 1987, Diaz and Rosenberg 1995), and the effective loss of habitat and fauna as a result of hypoxia can have profound effects on ecosystem energetics (Diaz and Rosenberg 2008).

Estimates of biomass lost due to hypoxia are approximately 10,000 megatons of carbon per year, or 5% of the Bay's total secondary production. Under healthy circumstances, an estimated 60% of benthic energy would be passed up the food chain in the Bay; therefore, it is estimated that hypoxic conditions in the Bay result in 6,000 megatons of carbon being lost as food energy for fisheries (Diaz and Rosenberg 2008).

Science and Management Actions (to Date and Planned)

Eutrophication and its influence on hypoxia have been studied extensively in the Chesapeake Bay in an effort to produce information useful for effective management of water quality and critical habitats (Boesch et al. 2001, Kemp et al. 2005). For example, Bay research has demonstrated that nitrogen and phosphorus limitation for phytoplankton growth vary seasonally and regionally (Fisher et al. 1999, Malone et al. 1986), underscoring the need to regulate both nitrogen and phosphorus inputs to the Bay. In addition, studies in this system have shown that nitrogen removal from Chesapeake Bay through denitrification appears to be inhibited compared with estimates for other coastal ecosystems that do not experience seasonal hypoxia (Kemp et al. 1990). The Chesapeake Bay has also been subjected to long-term water quality monitoring, as well as a long-term benthic monitoring program that has been a key metric for assessing progress towards nutrient- and hypoxia-related goals.

Historically, nutrient dynamics within the watershed have received less attention than in the Bay proper (Boesch et al. 2001), but there has recently been an increasing focus on the watershed processes. For example, it is clear that riparian zones, even in urban areas, are 'hotspots' of ecological function and can be effective at sequestering nutrients from runoff (Groffman et al. 2003). However, riparian zones can contribute to nitrogen loading in the Bay watershed when they undergo the "urban stream syndrome" (Groffman et al. 2004), which occurs where impermeable surfaces increase runoff leading to channel incision and lower water tables, thereby reducing denitrification potential in the watershed. This information has contributed to ongoing water quality improvement efforts.

A reduction in eutrophication has been a top priority for management of the Chesapeake Bay in the last few decades (Boesch et al. 2001). Scientific research has improved public awareness and political interest in reversing eutrophication in the Chesapeake Bay (Malone et al. 1993), and a series of policies and adaptive management plans have evolved (Boesch et al. 2001). The Chesapeake Bay became the first estuary in the United States to be targeted by Congress for restoration and protection. An outgrowth of this recognition was the formation of a regional partnership of Bay states, the District of Columbia, and Federal agencies. An agreement to cooperatively work together to protect the Bay was codified in The Chesapeake Bay Agreement of 1983 (http://www.chesapeakebay.net/content/ publications/ cbp_12512. pdf). This agreement was signed by what would become the Chesapeake Executive Council, a group comprised of the governors of Maryland, Virginia, and Pennsylvania; the mayor of the District of Columbia; the administrator of the EPA; and the chair of the Chesapeake Bay Commission.

In support of this agreement and the Chesapeake Bay Program, a Scientific and Technical Advisory Committee was formed to provide the most recent and accurate information. In 1987, a second agreement was signed, pledging to reduce nitrogen and phosphorus inputs into the Bay by 40% by the year 2000. In 1992, amendments were added, reaffirming the original goals of the 1987 agreement, but also pledging the development of tributary-specific strategies to reduce nutrient inputs.

Recognizing a lack of progress towards restoration of the Bay, members of the Chesapeake Bay Program drafted and signed the Chesapeake 2000 Agreement (C2K; http://www.chesapeakebay.net/ content/publications /cbp _12081.pdf), which provided comprehensive and specific direction for improving the water quality in the Bay, particularly related to nutrients. Recognizing the need to address Bay water quality within the watershed, the governors of Delaware, New York, and West Virginia entered into an agreement with the Chesapeake Executive Council to meet the goals of the C2K. This agreement was more aggressive than the previous one and called for, based on 1985 levels, a 48% reduction in nitrogen and 53% in phosphorus inputs. In 2003 and 2004, 36 tributary strategies were completed that outlined specific measures for each tributary to reduce the inputs of nutrients into the Bay. A history of Chesapeake Bay eutrophication and the evolution of public policy and awareness are available in Boesch et al. (2001).

Even more recently, in May 2009, President Obama issued Executive Order 13508 on Chesapeake Bay Protection and Restoration. To bring the full weight of the federal government to address the Chesapeake's challenges, the Executive Order established the Federal Leadership Committee (FLC) for the Chesapeake Bay, which is chaired by the Administrator of the US Environmental Protection Agency and includes senior representatives from the departments of Agriculture, Commerce, Defense, Homeland Security, Interior and Transportation. The EO charged the FLC with developing and implementing a new strategy for protection and restoration of the Chesapeake region. One of the primary strategies of the EO is "to restore clean water" in the bay, with the specific goal of reducing nitrogen, phosphorus, sediment and other pollutants to meet Bay water quality goals for dissolved oxygen, clarity, chlorophyll-a and toxic contaminants (http://executiveorder.chesapeakebay. net/).

Research on nutrient cycling in soil-water systems of the Chesapeake Bay watershed provides fundamental insight into the movement of nutrients in agricultural systems and their impact on water resources. This research has led to the development of models for better predictions of nutrient transport via runoff as well as practical tools, such as the "Phosphorus Index" (adopted by all states in the Bay watershed), to guide nutrient management decisions on agricultural fields.

The advent of Comprehensive Nutrient Management Planning has resulted in the development and implementation of best management practices (BMPs) for water and air quality protection. These BMPs range from traditional conservation practices—such as cover crops, riparian buffers, and constructed wetlands—to new technologies for precision application of fertilizers and manures to new practices for controlling and filtering drainage waters. Some of these BMPs can be transferred to nonagricultural uses, and ongoing research promises to deliver another generation of BMPs that immobilize nutrients and sediment.

A recent assessment of water quality trends in rivers feeding the Bay showed significant improvements in loadings of nitrogen (72% of sites showed downward trends), total phosphorus (81% of sites), and sediment (43% of sites), indicating that management actions are having some effect in reducing nutrients and sediments (Langland et al. 2006).

However, to date, lower nutrient input has not improved dissolved oxygen levels overall in the Chesapeake Bay, although it has caused small-scale reversals in hypoxia (Diaz and Rosenburg 2008).

Clearly, there are complex and poorly understood mechanisms that are acting to delay recovery of some ecosystem components (e.g., seagrass beds), and these mechanisms are priorities for scientific attention. Restoration of seagrass, oyster, and marsh habitats are expected to help the Bay's recovery from eutrophication and hypoxia by priming key ecological processes that will enhance recovery through biological positive-feedback mechanisms (Kemp et al. 2005).

Future Outlook

Small improvements in dissolved oxygen conditions are expected for the mainstem of the Chesapeake Bay, and the outlook for rivers draining into the Bay ranges from slight deterioration to slight improvement (Bricker et al. 2007).

This projection is an improvement from the last eutrophication outlook presented in 1999 (Bricker et al. 1999), when the Bay and its tributaries were expected to experience small to large levels of deterioration.

These projections are based on the prediction of expected nutrient load increases from wastewater treatment, septic tanks, agriculture, and urban runoff.

Current potential management concerns include increasing nutrient loads from sources such as wastewater treatment, agriculture, urban runoff, atmospheric deposition, on-site septic tanks, and combined sewer overflow, particularly because an increase in population is expected in the watershed.

To mitigate these potential problems, Bricker et al. (2007) suggests changes in land use policies that could limit urban sprawl and concentrate development, which could lead to more efficient treatment of runoff and waste.

References

Boesch DF, Brinsfield RB, Magnien RE. 2001. Chesapeake Bay eutrophication: Scientific understanding, ecosystem restoration, and challenges for agriculture. *J. Environ. Qual.* 30: 303-320.

Boicourt WC, Kuzmic M, Hopkins TS. 1999. The inland sea: circulation of Chesapeake Bay and the Northern Adriatic. In T. Malone, A. Malej, L. Harding, N. Smodlaka, R. Turner (eds.) Ecosystems at the land-sea margin: drainage basin to coastal sea. American Geophysical Union, Washington DC. P. 81-129.

Boicourt WC. 1992. Influences of circulation processes on dissolved oxygen in the Chesapeake Bay. In: Smith DE, Leffler M, Mackiernan G (eds.) Oxygen dynamics in the Chesapeake Bay: a synthesis of recent researchy. Maryland Sea Grant Publication, College Park, MD, p 7-59.

Boynton WR, Garber JH, Summers R, Kemp WM. 1995. Inputs, transformations, and transport of nitrogen and phosphorus in Chesapeake Bay and selected tributaries. *Estuaries* 18: 285-314.

Boynton WR, Kemp WM. 2000. Influence of river flow and nutrient loading on selected ecosystem processes and properties in Chesapeake Bay. In: Hobbie J. (ed.) Estuarine science: a synthetic approach to research and practice. Island Press, Washington, DC, p. 269-298.

Bricker SB, Clement CG, Pirhalla DE, Orlando SP, Farrow DRG. 1999. National Estuarine Eutrophication Assessment: effects of nutrient enrichment in the nation's estuaries. NOAA, National Ocean Service, *Centers for Coastal Ocean Science*, Silver Spring, MD.

Bricker S, Longstaff B, Dennison W, Jones A, Boicourt K, Wicks C, Woerner J. 2007. Effects of nutrient enrichment in the Nation's Estuaries: A decade of change. NOAA Coastal Ocean Program Decision Analysis Series No. 26. *National Centers for Coastal Ocean Science*, Silver Spring, MD. 322 pp.

Brush GS. 1984. Stratigraphic evidence of eutrophication in an estuary. Water Resour. Res. 20: 531-541.

Caddy JF. 1993. Toward a comparative evaluation of human impacts on fishery ecosystems of enclosed and semienclosed seas. *Rev. Fish Sci.* 1: 57-95.

Cohen RRH, Dresler PV, Philips EJP, Cory RL. 1984. The effect of the Asiatic clam Corbidula fulminea on phytoplankton of the Potomac River, *Maryland. Limnol. Oceanogr.* 29:170-18.

Cooper SR, Brush GS. 1991. Long-term history of Chesapeake Bay anoxia. *Science* 254: 992-996.

Cooper SR, Brush GS. 1993. A 2500-year history of anoxia and eutrophication in Chesapeake Bay. *Estuaries* 16: 617-626.

Diaz RJ, Rosenberg R. 1995. Marine benthic hypoxia: a refvfiew of its ecological effects and the behavioral responses of benthic macrofauna. Oceanogr. *Mar. Biol. Annu. Rev.* 33:245-303.

Diaz RJ, Rosenberg R. 2008. Spreading dead zones and consequences for marine ecosystems. *Science* 321:926- 929.

Fisher TR, Gustafson AB, Sellner K, Lacuture R, Haas LW, Magnien R, Karrh R, Michael B. 1999. Spatial and temporal variation in resource limitation in Chesapeake Bay. *Mar. Biol.* 133: 763-778.

Hagy JD, Boynton WR, Keefe CW, Wood KV. 2004. Hypoxia in Chesapeake Bay, 1950-2001: Long-term change in relation to nutrient loading and river flow. *Estuaries* 27: 634-658.

Holland AF, Shaughnessy AT, Hiegel MH. 1987. Long-term variation in mesohaline Chesapeake Bay macrobenthos: spatial and temporal patterns. *Estuaries* 10: 227-245.

Groffman PM, Crawford MK. 2003. Denitrification potential in urban riparian zones. *Journ. Environ. Qual.* 32:1144- 1149.

Groffman PM, Law NL, Belt KT, Band LE, Fisher GT. 2004. Nitrogen fluxes and retention in urban watershed ecosystems. *Ecosystems* 7: 398-403.

Karlsen AW, Cronin TM, Ishman ES, Willard DA, Holmes CW, Marot M, Kerhin R. 2000. Historical trends in Chesapeake Bay dissolved osygen based on benthic Foraminifera from sediment cores. *Estuaries* 23: 488-508.

Kemp WM, Sampou P, Caffrey J, Mayer M, Henriksen K, Boynton WR. 1990. Ammonium recycling versus denitrification in Chesapeake Bay sediments. *Limnol. Oceanogr.* 35: 1545-1536.

Kemp WM, Boynton WR, Adolf JE, Boesch DF, Boicourt WC, Brush G, Cornwell JC, Fisher TR, Glibert PM, Hagy JD, Harding LW, Houde ED, Kimmel DG, Miller WD, Newell RIE, Roman MR, Smith EM, Stevenson JC. 2005. Eutrophication of Chesapeake Bay: Historical trends and ecological interactions. *Mar. Ecol. Prog. Ser.* 303: 1-29.

Langland MJ, Raffensperger JP, Moyer DL, Landwehr JM, Schwarz GE. 2006. Changes in Streamflow and Water Quality in Selected Nontidal Basins in the Chesapeake Bay Watershed, 1985-2004. Scientific Investigations Report 2006–5178 US Department of the Interior, *US Geological Survey*. 74p.

Malone T, Kemp WM, Ducklow H, Boynton W, Tuttle J, Jonas R. 1986. Lateral variation in the production and fate of phytoplankton in a partially stratified estuary. *Mar. Ecol. Prog. Ser.* 32: 149-160.

Malone T, Boynton W, Horton T, Stevenson C. 1993. Nutrient loading to surface waters: Chesapeake case study. In: Uman MF (ed.) Keeping pace with science and engineering. National Academy Press, Washington, DC p. 8-38.

Newell RIE. 1988. Ecological changes in Chesapeake Bay: Are they the result of overharvesting the Eastern oyster (Crassostrea virginica)? In: Lynch MP, Krome EC (eds.) Understanding the estuary: advances in Chesapeake Bay research. Chesapeake Research Consortium Publication 129 (CBP/TRS 24/88) Bloucester Point, VA. pp. 536-546.

Newell RIE, Koch EW. 2004. Modeling seagrass density and distribution in response tochanges in turbidity from bivalve filtration and seagrass sediment stabilization. *Estuaries* 27: 793-806.

Niklitschek EJ. 2001. Bioenergetics modeling and assessment of suitable habitat for juvenile Atlantic and Shortnose stugeons (Acipenser oxyrinchus and A. brevirostrum) in the Chesapeake Bay. PhD. Dissertation, University of Maryland, College Park, MD.

Zimmerman AR, Canuel EA. 2002. Sediment geochemical records of eutrophication in the mesohaline Chesapeake Bay. *Limn. Oceanogr.* 47: 1084-1093.

Pensacola Bay

Physical Description of the System

The Pensacola Bay system is a network of estuarine bays located in the far western panhandle of Florida. The Bay system is arranged in two major arms that combine to form Pensacola Bay proper. The component bays include Escambia Bay, Blackwater Bay, East Bay, and Pensacola Bay (Figure 1). Santa Rosa Sound is relatively distinct shallow back-bay that joins the lower reach of Pensacola Bay and separates Santa Rosa Island from the mainland eastward to Destin, Florida, and Choctawhatchee Bay, Florida. The combined Pensacola Bay system, excluding Santa Rosa Sound, is medium sized (370 km^2) and shallow (mean depth is 3 m). Tides occur once per day and are relatively small, 15 to 65 centimeters.

Three major watersheds drain into the Bay via the Escambia, Blackwater, and Yellow Rivers (Figure 1).

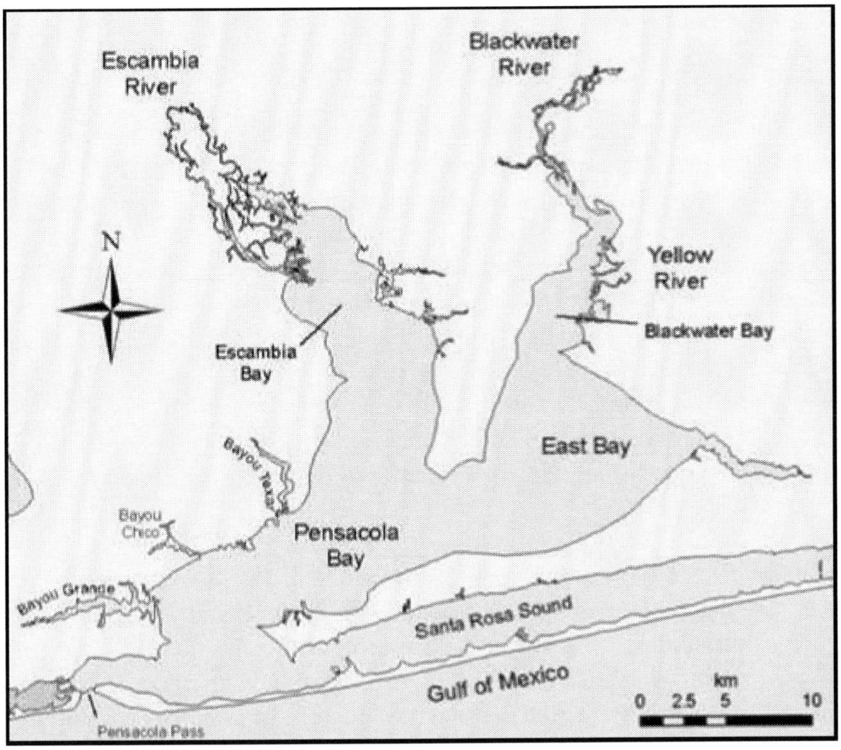

Figure 1. Map of Pensacola Bay, Florida.

History of Hypoxia (Issue, Causes, Economic and Ecosystem Impacts)

The condition of the Pensacola Bay system, particularly that of Escambia Bay, became a matter of public concern as early as the late 1960s after significant industrial point source discharges began and extensive fish kills and hypoxia were observed (Olinger et al. 1975).

The large fish kills prompted actions that ultimately led to elimination of the point source loadings by the mid-1970s. Olinger et al. (1975) provides a remarkable early compilation of ecological conditions in the Bay, intended principally to document the recovery of the system following reductions in industrial waste loads. These reports indicate that hypoxia was present in the Bays in the early 1970s, but does not provide enough data to evaluate the average extent or severity.

EPA monitored the Bay quarterly from 1996-2001 (US EPA 2005) and monthly from 2002- 2004 (Hagy and Murrell 2007). Whereas differences in survey methodology and data reporting mostly preclude quantitative analysis of ecological changes since the early 1970s, comparison with the most recent data suggests that neither recovery nor further degradation has occurred. The major ecological concerns of the early 1970s continue: bottom water hypoxia, loss of seagrass habitats, toxic contamination, and degradation of biotic communities (Hagy et al. 2008).

Hypoxia occurs frequently and extensively in bottom waters of Pensacola Bay (Hagy and Murrell 2007). An average of 25% of the Bay bottom is affected during the summer. Moderate river flow conditions create the highest potential for hypoxia in the Bay. In 2004, nearly 40% of the Bay bottom was hypoxic for two consecutive months (Hagy et al. 2008). Strong winds can bring hypoxic water onto shallow shoals, giving hypoxia the potential to degrade nearly all habitats in the Bay. Hypoxia has not been observed in the lower Bay, south and west of Bayou Texar, where tidal exchange with the Gulf of Mexico appears to be adequate to prevent hypoxia.

The causes of hypoxia in Pensacola Bay include strong salinity stratification, low tidal mixing energy, and sluggish estuarine circulation, which create an optimal physical environment for hypoxia (Hagy and Murrell 2007). On the other hand, water clarity in the Bay is often relatively high, chlorophyll-*a* concentrations are low to moderate, and phytoplankton production is moderate (Murrell et al. 2007). Correspondingly, rates of oxygen consumption in the water and sediments are relatively low (Murrell et al. 2009). These reflect low to moderate rates of nutrient loading, which can be related to low population density and minimal nutrient-intensive uses of much of the watershed (Hagy et al. 2008). The simultaneously low level of anthropogenic nutrient enrichment and eutrophication and high degree of hypoxia and seagrass loss suggests that the Bay has not been able to recover from earlier impacts.

The impact of hypoxia on biological communities in Pensacola Bay has not been well documented, but one can infer from the extent, frequency, and severity of hypoxia that affected habitats are almost certainly in a poor ecological condition. Limited data from the Florida Inshore Monitoring and Assessment Program show that the numbers of benthic macrofauna in Pensacola Bay in 2003 were only 5 to 10% of numbers in healthier regions of the Florida west coast. Livingston (1999) found that substantial biomass of infaunal animals occurred in only a narrow zone around the perimeter of the Bay, beyond the reach of persistent hypoxia. It has also been shown that chronic exposure to hypoxia imposed physiological stress on fish in Pensacola Bay, preventing them from reproducing (Thomas et al. 2007). Although the massive seagrass loss in Pensacola Bay has not been linked directly to hypoxia, it could reduce the abundance of animals that graze on algae growing on seagrasses, thus inhibiting their recovery. For example, grass shrimp enhanced growth of *Ruppia maritima*, a seagrass that was once abundant in Pensacola Bay (McCall and Rakocinski 2007). Conceivably, the massive loss of seagrasses that has occurred and the extensive recurrent hypoxia could be mutually reinforcing, promoting persistence of these conditions despite currently unremarkable rates of nutrient loading.

Science and Management Actions (to Date and Planned)

The record of scientific research addressing water quality and ecological condition in Pensacola Bay began in earnest with the Olinger et al. (1975) studies. These studies documented many of the key features of the ecology of the Bay, including the incidence of hypoxia and some of its physical causes. Papers by EPA scientists, mostly since 2000, provide more detail on many aspects of the physical and biological conditions that control hypoxia in the Bay (see Hagy et al. 2008). EPA will continue studies of Pensacola Bay as part of a regional research program in the northeast Gulf of Mexico. The project will focus on providing data and methods to develop numeric water quality criteria for nutrients and nutrient-related water quality parameters.

Because Pensacola Bay is a relatively small estuary, no management programs have been created specifically to manage and improve its water quality (e.g., such as those created for Chesapeake Bay or the Gulf of Mexico). Florida's Impaired Waters Rule provides some numeric guidelines for listing estuaries as impaired for chlorophyll-*a* and dissolved oxygen, but there are no enforceable water quality standards for nutrients or dissolved oxygen for the Bay.

On the other hand, the state of Florida has recently increased its focus on developing water quality criteria for nutrients in its estuaries and coastal waters. Portions of Escambia Bay have been listed as impaired for an excessive increase in chlorophyll-*a*. Establishing levels for total maximum daily loads has been proposed and would call for significant reductions in both nitrogen and phosphorus loading to upper Escambia Bay. These actions signal an increase in regulatory attention to nutrients in Pensacola Bay. A continuation of scientific research by EPA and others could provide an improved scientific basis for these actions. Presently, the best hope for improved management of water quality in the Bay seems to be the likely prospect that the State of Florida will soon develop and adopt numeric nutrient criteria for Pensacola Bay. EPA is collaborating with the Florida Department of Environmental Protection to support its efforts to achieve this objective for Pensacola Bay and other estuaries in Florida.

Future Outlook

The relatively undeveloped status of a large portion of the watershed of Pensacola Bay is one of its best assets. Arguably, low population density in the watershed has prevented eutrophication and hypoxia from becoming much worse. However, the population of the coastal counties surrounding the Bay, especially Santa Rosa County, Florida, is projected to grow nearly 3.5-fold by 2060 (Zwick and Carr 2006), most likely increasing nutrient loading to this sensitive system. A regulatory approach utilizing the authority of the Clean Water Act and based on numeric nutrient criteria will be the best way of ensuring that this growth occurs in a manner that does not further degrade the Bay.

Although the available research clearly indicates that increased nutrients could harm water quality in the Bay, it is less clear that nutrient reductions alone will be sufficient to restore the Bay. A more active restoration program may be needed. Research is needed to determine the best methods for restoring ecological function in the Bay in order to simultaneously increase the extent of seagrass habitats, reduce hypoxia, and promote better biological condition overall.

References

Hagy JD III, Murrell MC. 2007. Susceptibility of a Gulf of Mexico estuary to hypoxia: An analysis using box models. Estuarine, *Coastal and Shelf Science*. 74: 239-253.

Hagy JD III, Kurtz JC, Greene RM. 2008. An approach for developing numeric nutrient criteria for a Gulf coast estuary. US Environmental Protection Agency, Office of Research and Development, National Health and Environmental Effects Research Laboratory, Research Triangle Park, NC. EPA/600/R-08/004. 48 pp.

Livingston RJ. 1999. Ecology and Trophic Organization, Section 5a, Volume 4 in Pensacola Bay System Environmental Study, unpublished report from Champpion International Corporation, 125 pp.

McCall DD, Rakocinski CF. 2007. Grass Shrimp (Palaemonetes spp.) play a pivotal trophic role inenhancing Ruppia maritima. *Ecology* 88(3): 618-624.

Murrell M C, Hagy JD III, Lores EM, Greene RM. 2007. Phytoplankton production in relation to nutrient distributions in a subtropical estuary: importance of freshwater flow. *Estuaries and Coasts.* 30(3): 390-402.

Murrell MC, Campbell JG, Hagy JD III, Caffrey J. 2009. Effects of irradiance on benthic and water column processes in a shallow micro-tidal estuary: Pensacola Bay, Florida, USA. Estuarine, Coastal and Shelf Science 81: 501-512.

Olinger LW, Rogers RG, Fore PL, Todd RL, Mullins BL, Bisterfield FT, Wise LA. 1975. Environmental and recovery studies of Escambia Bay and the Pensacola Bay system, Florida. Report No. 904/9-76-016, US Environmental Protection Agency, Atlanta, Georgia. 468 pp.

Thomas P. Saydur Rahman M, Khan IA, Kummer JA. 2007. Widespread endocrine diisruption and reproductive impainment in an estuarine fish population exposed to seasonal hypoxia. *Proceedings of the Royal Society* B. 274: 2693-2701.

US EPA. 2005. The Ecological Condition of the Pensacola Bay System, Northwest Florida (1994-2001). US Environmental Protection Agency, Office of Research and Development, National Health and Environmental Effects Research Laboratory, Gulf Ecology Division, Gulf Breeze, FL. EPA/620/R-05/002.

Zwick PD, Carr MH. 2006. Florida 2060, A population distribution scenario for the State of Florida. Geoplan Center, University of Florida, Gainsville, FL, USA. 29 pp.

NORTHERN GULF OF MEXICO

Physical Description of the System

The largest zone of oxygen-depleted coastal waters in the United States and arguably the second largest on Earth is in the northern Gulf of Mexico on the Louisiana continental shelf (Figure 1). The Gulf hypoxic zone typically occurs in waters 5-60 m in depth up to 125 km offshore from west of the Mississippi River Delta and occasionally extending to the upper Texas coast.

The dominant sources of freshwater, sediments, and nutrients to the northern Gulf of Mexico are the Mississippi and Atchafalaya Rivers. The watershed for these rivers (commonly referred to as the Mississippi Atchafalaya River Basin, or MARB) encompasses 41% of the contiguous United States (Figure 2). The average annual streamflow delivered from the MARB to the Gulf of Mexico during the period from 1981 to 2005 was 21,700 cubic meters per second (m^3/s); the average annual flux of total nitrogen and total phosphorus during that period was 1.47 and 0.14 million metric tons, respectively (Aulenbach et al. 2007). Streamflow from the MARB enters the Gulf of Mexico through two deltas— about two thirds of the flow enters via the Mississippi River Birdfoot Delta (southeast of New Orleans, Louisiana), and one third via the Atchafalaya River Delta (200 km west on the central Louisiana coast). The freshwater discharge is carried predominantly westward along the Louisiana/Texas inner to mid-continental shelf, especially during peak spring discharge. This coastal ocean margin is characterized as a relatively shallow, open coastline with complex circulation patterns, weak tidal energies, generally high water temperatures, seasonally varying stratification strength, and large inputs of freshwater that effectively result in an unbounded estuary, stratified for much of the year. Water column density stratification, which is critical to bottom water oxygen depletion, is dominated by vertical salinity gradients, but thermal warming of surface waters intensifies summer stratification strength. Water column structure is also highly influenced by wind stress, frontal weather bands, hurricanes, and mixing of buoyant river plumes.

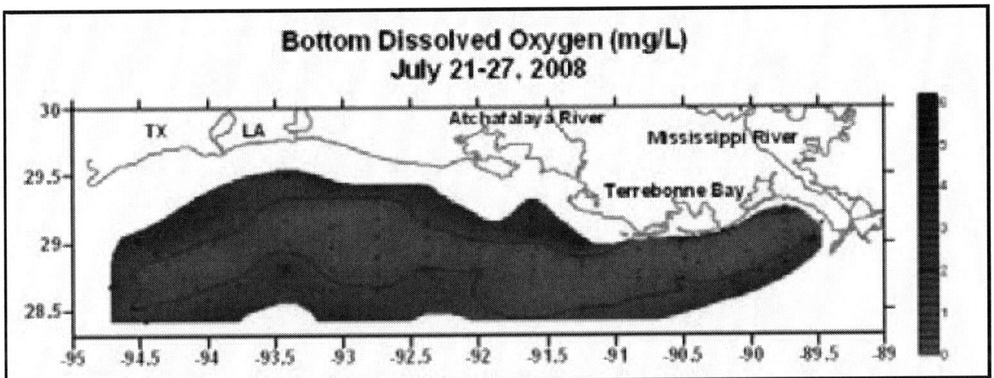

Data source: N. Rabalais, Louisiana Universities Marine Consortium, NOAA, map by B. Babin.

Figure 1. Map of bottom water oxygen levels in mg/L. Black line is less then 2 mg/L (hypoxic).

History of Hypoxia (Issue, Causes, Economic and Ecosystem Impacts)

Extensive bottom water hypoxia forms each year between May and September. Since 1985 (when systematic mapping was started), the hypoxic area has averaged 13,808 km^2 and achieved its maximum size in 2002 at 22,000 km^2 (Figure 3). While low dissolved oxygen is commonly considered a bottom-water condition, oxygen-depleted waters often extend up into the lower half to two-thirds of the water column. Longterm increases in nutrient loads from the MARB, coupled with water column stratification, have been implicated as the primary causes of hypoxia (CENR 2000, US EPA 2007). However, the complex physical, chemical, and biological processes along the coastal ocean margin—which consume, transform, and remineralize riverine nutrients and organic matter and ultimately result in bottom water oxygen depletion—remain poorly resolved in space and time.

The nutrients delivered with freshwater inputs support primary productivity within the immediate vicinity of the river discharges as well as across the broader continental shelf. The settling of fixed carbon to the lower water column and seabed in the form of senescent phytoplankton, zooplankton fecal pellets, or aggregates provides a large carbon source for decomposition by aerobic bacteria, which in turn leads to hypoxia. Tropical storms, hurricanes, and cold front passages disrupt the hypoxia until stratification reestablishes and oxygen depletion processes continue.

Low oxygen events on the Louisiana-Texas continental shelf have been reconstructed over the past centuries using the relative abundance of low oxygen tolerant benthic foraminifera in sediment cores (Osterman et al. 2005). These records show that low oxygen events have increased over the past 50 years. Additionally, regression model hindcasts using historical Mississippi River discharge and nitrate concentrations indicated that large-scale hypoxia has likely been present along the continental shelf since the mid-1950s (Greene et al. 2009). More recently, the areal extent of the hypoxic zone increased from an average of 6,900 km^2 during 1985-1992 to 15,570 km^2 during 2005-2009 (Rabalais and Turner 2006, LUMCON 2009) (Figure 3).

The increased prevalence of Gulf hypoxia over recent decades has been related to increases in nutrient loads (CENR 2000, US EPA 2007). However, it has been demonstrated that hypoxia events have occurred for centuries driven by high streamflow events that flush nutrients from wetland ecosystems and stratify ocean waters (Osterman et al. 2005, Swarzenski et al. 2008). Alterations of the Mississippi and Atchafalaya Rivers for transportation and flood control over two centuries have significantly lessened the assimilation of nutrients in the watershed and changed the pattern of freshwater discharge to the coastal ocean margin (US EPA 2007). Concurrent increases in anthropogenic inputs of nutrients to the watershed have contributed to increased eutrophication and hypoxia (from Bricker et al. 1998, Rabalais in Bricker et al. 2007).

The region supports some of the most valuable commercial and recreational fisheries in the United States (Diaz and Solow 1999, Chesney et al. 2000, Zimmerman and Nance 2001). For example, Texas and Louisiana lead all states in catches of shrimp, which is the largest US commercial fishery (NOAA 2007). However, partitioning the impacts of hypoxia on living resources from other ecosystem stressors (e.g., climate change and overfishing) has proven difficult and significant knowledge gaps remain. Fish kills are occasionally reported and significant impacts to benthic fauna have been well documented (e.g., Rabalais et al. 2001). Zimmerman and Nance (2001) suggest that severe hypoxic conditions may block the migration of shrimp from nearshore to offshore habitats.

Additionally, brown shrimp are subjected to a significant amount of habitat loss due to hypoxia (Craig et al. 2005), congregating in suboptimal environments along the hypoxic zone edge, possibly causing a reduction in growth (Craig and Crowder 2005). Recent advancements in biomarker techniques have suggested that hypoxia may be causing reproductive impacts in croaker as well (Thomas et al. 2007).

Science and Management Actions (to Date and Planned)

In 1998, the EPA established the Mississippi River/Gulf of Mexico Watershed Nutrient Task Force. The Task Force brought together Federal agencies, states, and tribes to consider options for reducing, mitigating, and controlling hypoxia in the northern Gulf of Mexico. In 1999 and 2000, the results of an integrated science assessment, requested by the Task Force and conducted under the auspices of the National Science and Technology Council, was published (CENR 2000, Brezonik et al. 1999, Diaz and Solow 1999, Doering et al. 1999, Goolsby et al. 1999, Mitsch et al. 1999, and Rabalais et al. 1999) (http://oceanservice.noaa.gov/products/pubs_hypox. html).

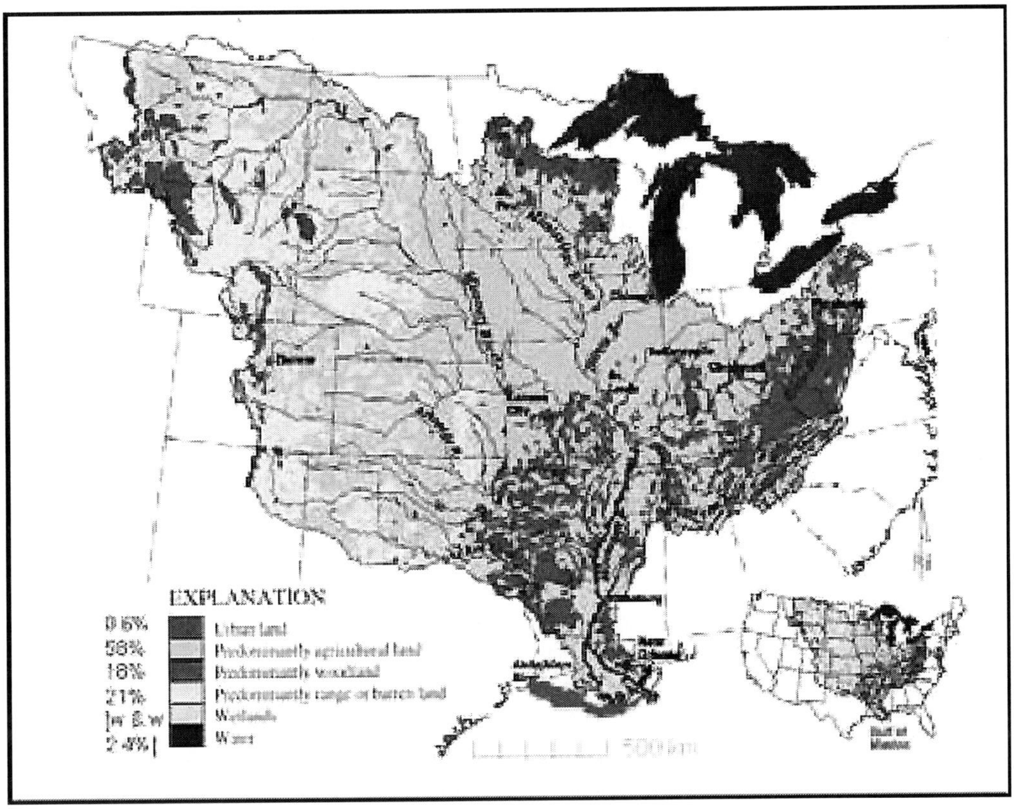

Figure 2. Mississippi River Watershed with Dead Zone shown in Red (modified from Mississippi River/Gulf of Mexico Watershed Nutrient Task Force 2004).

Using this science assessment, the Task Force published its first "Action Plan", entitled *Action Plan for Reducing, Mitigating, and Controlling Hypoxia in the Northern Gulf of Mexico* (Mississippi River/Gulf of Mexico Watershed Nutrient Task Force 2001), which was endorsed by Federal agencies, states, and tribal governments. The 2001 Action Plan coastal goal was to reduce the five-year running average size of the hypoxic zone to 5,000 km^2 by 2015 (Figure 3). It estimated that an overall reduction in nitrogen load of 30–45% would be required. The 2008 Plan identifies the requirement that phosphorus as well as nitrogen loads be reduced and increases the required reduction to at least 45%. The 2008 Plan also notes that reductions in nitrogen loads from 2001-2005 were from nitrogen forms other than nitrate, which is the primary form fueling spring primary production that leads to hypoxia. The Action Plan was based on a series of voluntary and incentive-based activities that address both reducing nutrient inputs and increasing assimilation in aquatic ecosystems, including proper timing and amount of fertilizer applications, best management practices on agricultural lands, restoration and creation of wetlands, river hydrology remediation, riparian buffer strips, nutrient removal from stormwater and wastewater, and coastal diversions.

In 2007, an updated science assessment was conducted by the EPA Science Advisory Board under the oversight of the Task Force (US EPA 2007). This update science assessment was used by the Task Force to develop the *Gulf Hypoxia Action Plan for Reducing, Mitigating, and Controlling Hypoxia in the Northern Gulf of Mexico and Improving Water Quality in the Mississippi River Basin* (Mississippi River/ Gulf of Mexico Watershed Nutrient Task Force 2008) (see Box 4 in Chapter 2). This plan retains the coastal goal of reducing the hypoxic zone to less than 5,000 km^2 by 2015, but understands the difficulty of meeting the goal. In this regard, the Task Force accepts the advice of the EPA Science Advisory Board on this issue: "The 5,000 km^2 target remains a reasonable endpoint for continued use in an adaptive management context; however, it may no longer be possible to achieve this goal by 2015... it is even more important to proceed in a directionally correct fashion to manage factors affecting hypoxia than to wait for greater precision in setting the goal for the size of the zone. Much can be learned by implementing management plans, documenting practices, and measuring their effects with appropriate monitoring programs" (US EPA 2007). The 2008 Action Plan, also like the 2001 Action Plan, calls for a reassessment after five years. The next reassessment will be conducted in 2013.

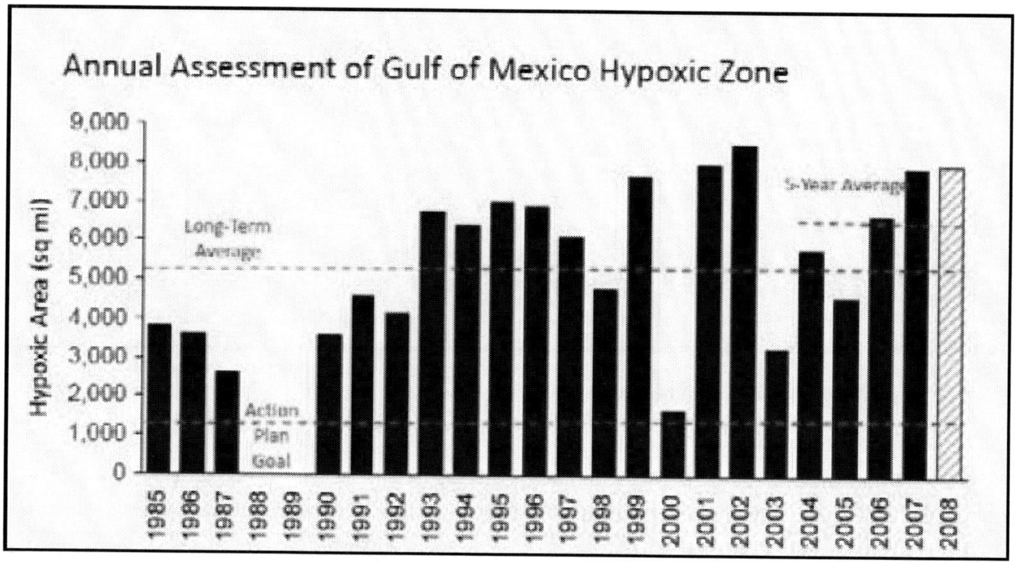

Figure 3. Northern Gulf of Mexico hypoxic area by year, including 5 year running and longterm averages. Action Plan Goal is 5000 km^2.

On June 16, 2010, USDA released a report, "Effects of Conservation Practices on Cultivated Cropland in the Upper Mississippi River Basin."(http://www.nrcs.usda.gov /technical/nri/ceap/umrb/index.html) This report, part of USDA's Conservation Effects Assessment Project (CEAP), showed that agricultural producers in the Upper Mississippi River Basin have had great success in reducing soil erosion through widespread adoption of sediment controlling conservation activities. The assessment also showed that more widespread adoption of effective nutrient control practices is needed to meet the challenges associated with nitrogen and phosphorous leaching and runoff. USDA has committed $320 million to improve water quality in priority watersheds through its Mississippi River Basin Healthy Watersheds Initiative (MRBI).

Future Outlook

Although overall total annual nutrient loads to the northern Gulf from 2001–2005 suggests a decline in nitrogen inputs relative to the previous 24-year average (Mississippi River/Gulf of Mexico Watershed Nutrient Task Force 2008), reductions in nitrate during the critical spring period have not occurred. Further, the five-year running average (2005–2009) of the hypoxic zone is 15,570 km^2, which is more than three times the Action Plan goal of 5,000 km^2. The adaptive management approach put forth in the Action Plans of 2008 and 2001 commits to a reliance on sound science and continual feedback between the interpretation of new information and improved management actions as the key to targeting actions within watersheds where they will be most effective. Significant variability in nutrient export rates, hypoxic zone size, and other parameters resulting from anthropogenic and/or climate change will make it difficult to assess the outcome of nutrient management actions. However, targeted monitoring on representative watersheds, continued monitoring of streamflow and nutrient flux, and research that addresses biogeochemical processes and improved model application will provide the most effective means of measuring results and providing feedback on performance.

References

Aulenbach BT, Buxton HT, Battaglin WA, Coupe RH. 2007. Streamflow and nutrient fluxes of the Mississippi-Atchafalaya River Basin and subbasins for the period of record through 2005: US Geological Survey Open-File Report 2007-1080. http://toxics.usgs.gov/pubs/of-2007-1080/index.html.

Brezonik PL, Bierman VJ, Alexander R, Anderson J, Barko J, Dortch M, Hatch L, Hitchcock GL, Keeney D, Mulla D, Smith V, Walker C, Whitledge T, Wiseman WJ. 1999. Effects of reducing nutrient loads to surface waters within the Mississippi River Basin and Gulf of Mexico. Topic 4 Report for the Integrated Assessment on Hypoxia in the Gulf of Mexico. NOAA Coastal Ocean Program Decision Analysis Series No. 18. Silver Spring, MD: *National Oceanic and Atmospheric Administration.*

Bricker S, Clement C, Frew S, Harmon M, Harris M, Pirhalla D. 1998. NOAA's Estuarine Eutrophication Survey. Volume 4: Gulf of Mexico Region. Silver Spring, MD. *Office of Ocean Resources Conservation Assessment.* 78 pp.

CENR (Committee on Environment and Natural Resources). 2000. Integrated Assessment of Hypoxia in the Northern Gulf of Mexico. National Science and Technology Council Committee on Environment and Natural Resources, Washington, D.C.

Chesney EJ, DM Baltz, RG Thomas. 2000. Louisiana estuarine and coastal fisheries and habitats: perspectives from a fish's eye view. Ecological Applications 10: 350-366.

Craig JK, Crowder LB. 2005. Hypoxia-induced habitat shifts and energetic consequences in Atlantic croaker and brown shrimp on the Gulf of Mexico shelf. Mar. Ecol. Prog. Ser. 294:79-94.

Craig JK, Crowder LB, Henwood TA. 2005. Spatial distribution of brown shrimp(Farfantepenaeus aztecus) on the northwestern Gulf of Mexico shelf: effects of abundance and hypoxia. Canadian Journal of Aquatic Science 62:1295-1308.

Diaz RJ, Solow A. 1999. Ecological and economic consequences of hypoxia—Topic 2, Gulf of Mexico hypoxia assessment: Silver Spring, MD, NOAA Coastal Ocean Program Decision Analysis Series, 86 pp., http:// oceanservice.noaa.gov/products/hypox_t2final.pdf.

Doering OC, Diaz-Hermelo F, Howard C, Heimlich R, Hitzhusen F, Kazmierczak R, Lee J, Libby L, Milon W, Prato T, Ribaudo M. 1999. Evaluation of economic costs and benefits of methods for reducing nutrient loads to the Gulf of Mexico. Topic 6 Report for the Integrated Assessment on Hypoxia in the Gulf of Mexico. NOAA Coastal Ocean Program Decision Analysis Series No. 20. Silver Spring, MD: National Oceanic and Atmospheric Administration.

Goolsby DA, Battaglin WA, Lawrence GB, Artz RS, Aulenbach BT, Hooper RP, Keeney DR, Stensland GS. 1999. Flux and sources of nutrients in the Mississippi-Atchafalaya River Basin. Topic 3 Report for the Integrated Assessment on Hypoxia in the Gulf of Mexico. NOAA Coastal Ocean Program Decision Analysis Series No. 17. Silver Spring, MD: National Oceanic and Atmospheric Administration.

Greene RM, Lehrter JC, Hagy JD III. 2009. Multiple regression models for hindcasting and forecasting midsummer hypoxia in the Gulf of Mexico. *Ecological Applications* 19(5): 1161-1175.

LUMCON. 2009. 'Dead Zone' again rivals record size: Press Release, Louisianna Universities Marine Consortium (LUMCON), http://www.gulfhypoxia.net/research/shelfwidecruises/2008/PressRelease08.pdf.

Mississippi River/Gulf of Mexico Watershed Nutrient Task Force. 2001. Action Plan for Reducing, Mitigating, and Controlling Hypoxia in the Northern Gulf of Mexico. Washington, D.C.: Offices of Wetlands, Oceans, and Watersheds, US Environmental Protection Agency.

Mississippi River/Gulf of Mexico Watershed Nutrient Task Force. 2004. A Science strategy to support management decisions related to hypoxia in the Northern Gulf of Mexico and excess nutrients in the Mississippi River Basin: Prepared by the Monitoring Modeling and Research Workgroup of the Mississippi River/Gulf of Mexico Watershed Nutrient Task Force, US Geological Survey Circular 1270, 58 pp.

Mississippi River/Gulf of Mexico Watershed Nutrient Task Force. 2008. Gulf Hypoxia Action Plan 2008 for Reducing, Mitigating, and Controlling Hypoxia in the Northern Gulf of Mexico and Improving Water Quality in the Mississippi River Basin. Washington, D.C.

Mitsch WJ, Day JW, Gilliam JW, Groffman PM, Hey DL, Randall GW, Wang N. 1999. Reducing nutrient loads, especially nitrate-nitrogen, to surface water, ground water, and the Gulf of Mexico. Topic 5 Report for the Integrated Assessment on Hypoxia in the Gulf of Mexico. NOAA Coastal Ocean Program Decision Analysis Series No. 19. Silver Spring, MD: National Oceanic and Atmospheric Administration.

NOAA. 2007. Nutrient Enhanced Coastal Ocean Productivity (NECOP) Program: National Oceanic and Atmospheric Administration, Washington, DC. http://www.aoml.noaa.gov/ocd/necop.

Osterman LE, Poore RZ, Swarzenski PW, Turner RE. 2005. Reconstructing a180-yr record of natural and anthropogenic induced hypoxia from the sediments of the Louisiana continental shelf: *Geology* 33(4): 329–332.

Rabalais NN, Turner RE, Justic D, Dortch Q, Wiseman WJ. 1999. Characterization of hypoxia. Topic 2 Report for the Integrated Assessment on Hypoxia in the Gulf of Mexico. NOAA Coastal Ocean Program Decision Analysis Series No. 15. Silver Spring, MD: National Oceanic and Atmospheric Administration.

Rabalais NN, Harper DE Jr., Turner RE. 2001. Response of nekton and demersal and benthic fauna to decreasing oxygen concentrations, p. 115-128. In: NN Rabalais and RE Turner (eds), Coastal Hypoxia Consequences for Living Resources and Ecosystems. Coastal and Estuarine Studies 58, American Geophysical Union, Washington D.C.

Rabalais NN, Turner RE. 2006. Oxygen depletion in the Gulf of Mexico adjacent to the Mississippi River, *in* Neretin, L.N., ed., Past and present marine water column anoxia: NATO Science Series, *IV-Earth and Environmental Sciences,* Kluwer, pp. 225–245.

Swarzenski PW, Campbell PL, Osterman LE, Poore RZ. 2008. A 1000-year sediment record of recurring hypoxia off the Mississippi River: the potnetial role of terrestrially-derived organic matter inputs. *Marine Chemistry* 109: 130- 142.

Thomas P, Rahman MS, Khan IA, Kummer JA. 2007. Widespread endocrine disruption and reproductive impairment in an estuarine fish population exposed to seasonal hypoxia. *Proc. Royal Society B-Biological Sciences* 274: 2693-2702.

US EPA. 2007. Hypoxia in the Northern Gulf of Mexico: An Update by the EPA Science Advisory Board: EPA-SAB08-004, December 2007, Washington, D.C., 275 pp.

Zimmerman AR, Nance JM. 2001, Effects of hypoxia on the shrimp industry of Louisiana and Texas, In Rabalais, NN, Turner RE (eds.), Coastal hypoxia—Consequences for living resources coastal and estuarine studies: Washington, D.C., American Geophysical Union, chap. 15, v. 58, pp. 293–310.

NORTHEAST PACIFIC CONTINENTAL SHELF (OREGON/WASHINGTON)

Physical Description of the System

The Oregon and Washington Shelf out to 200 meters water depth encompasses approximately 26,600 square kilometers (km^2) and has an average width of approximately 40 km. Generally, the shelf narrows from north to south, with a shelf width of more than 60 km off Washington and less than 20 km along southern Oregon. The exception to this general trend is a region called Heceta Bank in the central region of Oregon (43.8° to 44.6° N) which extends out to 65 km.

The shelf is part of the Northern California Current system. This system has strong seasonal variability, with northerly (from the north) winds dominating from April to October, and southerly winds dominating through the winter. The strong northerly winds induce coastal upwelling which brings nutrient-rich, low-oxygen waters onto the shelf (Figure 1).

The major source of terrestrial input to the system is from the Columbia River, with an annual mean outflow of 6484 cubic meters per second (m^3/s, 15 year mean from 1992 – 2006). River outflow tends to be low during the summer growing season and the input of nutrients from the river is negligible compared to contributions from upwelled waters (Hickey and Banas 2003). Thus, hypoxia off Washington and Oregon is unique in that it is due to natural sources rather than anthropogenic sources.

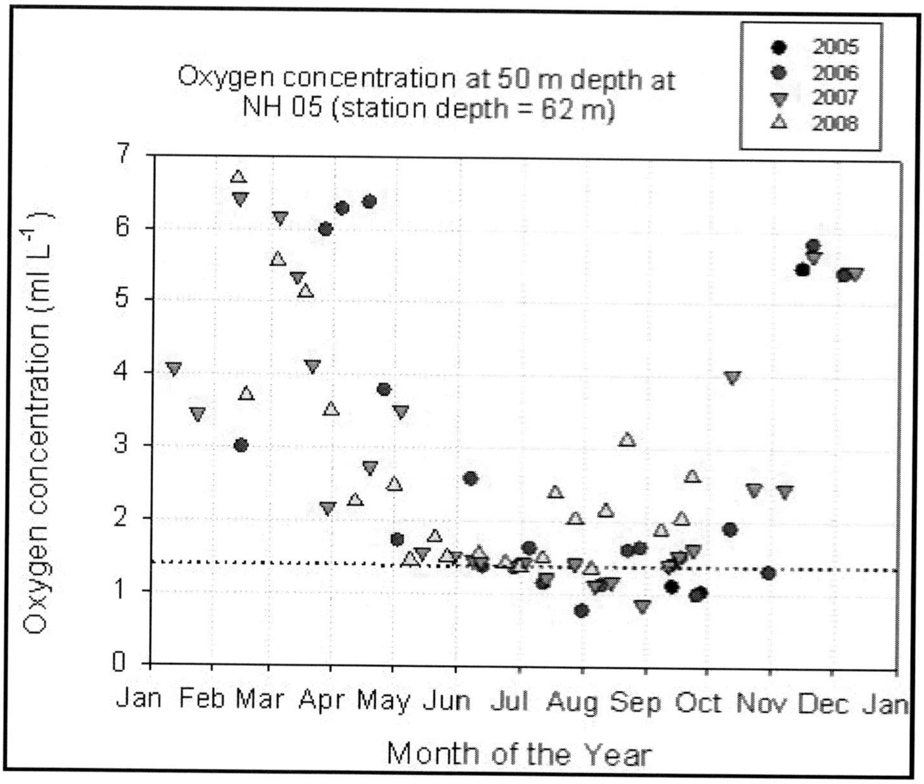

Figure 1. Seasonal change in bottom water oxygen concentration at a station 5 miles from shore off Newport, OR. The horizontal line at 1.4 ml/L represents hypoxic waters. Note that hypoxic waters appear only during the upwelling season (May-October). Source: Bill Peterson, NOAA NWFSC.

Economically, upwelling-induced productivity in Oregon-Washington Shelf waters supports a $50 million fishery for dungeness crabs. This is a pot fishery, so crabs trapped in the pot fishing gear are killed during summertime hypoxic events. The value of the salmon fishery is on the order of $10-20 million; however, the impact of hypoxia on the landings of the crabs and salmon has not yet been evaluated economically.

History of Hypoxia (Issue, Causes, Economic and Ecosystem Impacts)

Evidence of hypoxia on the Oregon-Washington continental shelf indicates that it is a seasonal occurrence (Figure 1) that has been present since at least the 1960s (Figure 2). However, there has not been a consistent sampling program developed to assess the degree and magnitude of hypoxia over time (note the lack of any data between 1970 and 1998 in Figure 2).

Data have been collected three to four times per year since 1998 during broad-scale surveys of zooplankton and salmon and biweekly since late 2005 (Figure 1).

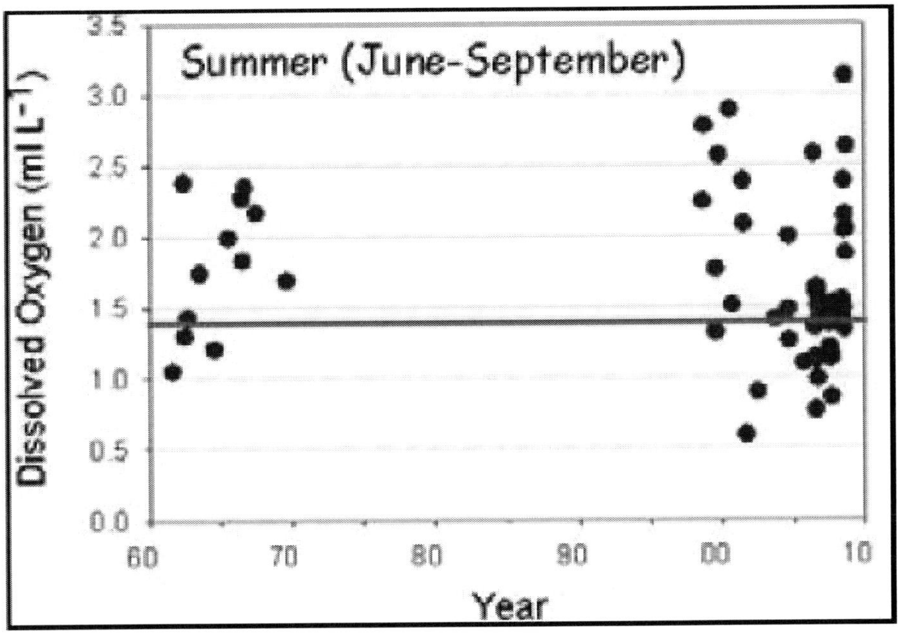

Figure 2. Oxygen concentration (ml/L) within 10m of the bottom at a station 5 miles offshore of Newport, OR. The thick, red line indicates hypoxic waters (1.4 ml/L). Although hypoxia was observed in the 1960s, it has been observed more frequently in this century. Source: Bill Peterson, NOAA NWFSC.

Cross-shelf transects (Figure 3) show that the hypoxic bottom waters can extend at least 20 to 30 meters off the bottom and occupy up to 30% of the water column. Low oxygen water moves onto the shelf in early spring (April or May) during the onset of upwelling. Throughout the summer, the oxygen level of bottom waters is reduced through the biological degradation of organic matter.

Wider regions of the shelf, where circulation patterns retain water for longer periods of time, tend to have more persistent and severe oxygen depletion (Figure 4).

In September 2007, 8,600 km^2 of the shelf had hypoxic bottom waters, covering 63% of the shelf area surveyed (Figure 4), making the Oregon-Washington shelf the second largest hypoxic region in the continental United States, second only to the Gulf of Mexico (which averages roughly 14,000 km^2).

Severe hypoxia events were observed in 2002 (Grantham et al. 2004) and 2006 (Chan et al. 2008). From these events, it is clear that in some areas, such as Heceta Bank (just south of Newport), oxygen concentrations fall to very low values and can persist through much of the summer, ultimately killing or displacing nearly all of the bottom-dwelling organisms.

Recent work on the impacts of hypoxic waters on the development of copepods indicate that oxygen concentrations below 1 ml/L (or 1.43 mg/L) greatly reduces the hatching rate of copepod eggs (Peterson unpublished). In instances where copepods are not able to move out of hypoxic zones, the reduced hatching rate may also be accompanied by reduced egg production (Marcus 2001, Sedlacek and Marcus 2005), severely impacting recruitment of species critical to the food web of many commercially important fish species.

The cause of the hypoxia is driven primarily by an interaction between circulation and biological activity. The typical upwelling season extends from April to October. During this time, periods of strong winds from the north bring nutrient-rich, low-oxygen waters onto the shelf. The nutrient-rich waters are mixed into the surface layer and fuel a highly productive planktonic food chain.

However, vertical transport of fecal pellets of grazers and sinking of nutrient-depleted phytoplankton blooms results in the degradation of massive amounts of organic matter on the sea floor.

Depending on the vertical flux and on the retention time of the water mass, the oxygen consumption can rapidly lead to hypoxic and even anoxic conditions.

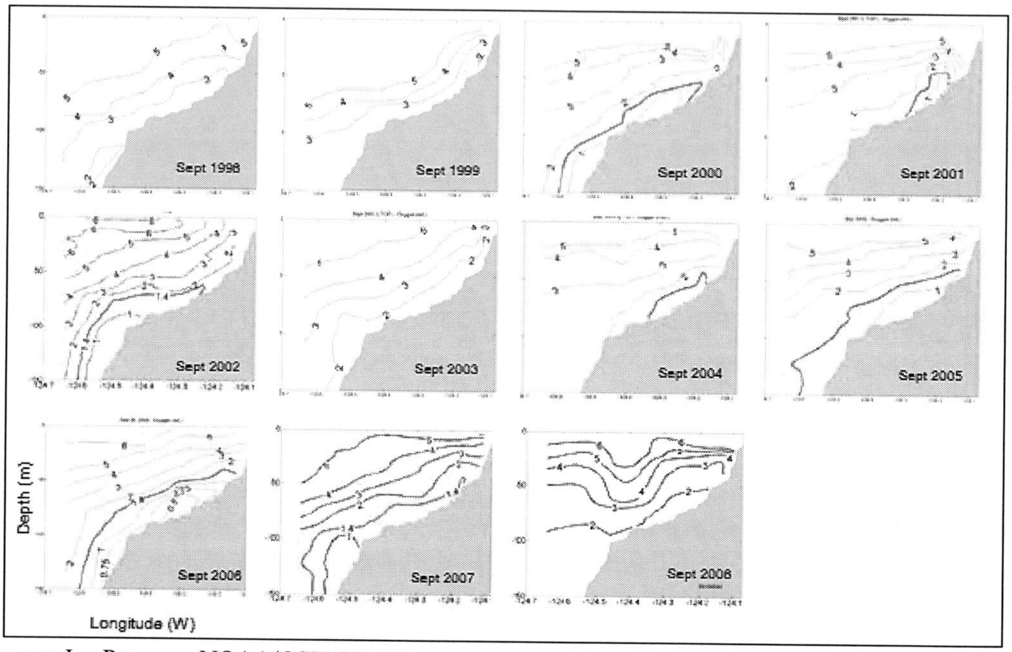

Source: Jay Peterson, NOAA/OSU CIMRS.

Figure 3. Oxygen concentration along a 46 km (25 mile) transect across the Oregon shelf (NH Line off Newport, Oregon). The thick, red contour indicates the region of hypoxia (1.4 ml/L). The gray shaded area is the bottom bathymetry.

Science and Management Actions (to Date and Planned)

To date, no specific scientific programs have been established to monitor the status of hypoxic zones along the Oregon-Washington shelf. The development of hypoxic regions along the coast is primarily driven by naturally occurring physical and biological processes.

Changes in the amount of nutrients upwelled to surface waters, water temperatures, and retention of water masses on the shelf likely contribute to the persistence, size, and severity of hypoxic zones.

The low oxygen zones have been shown to greatly impact bottom-dwelling fishes, shellfish, and other organisms inhabiting the shallow banks (i.e., Heceta Bank).

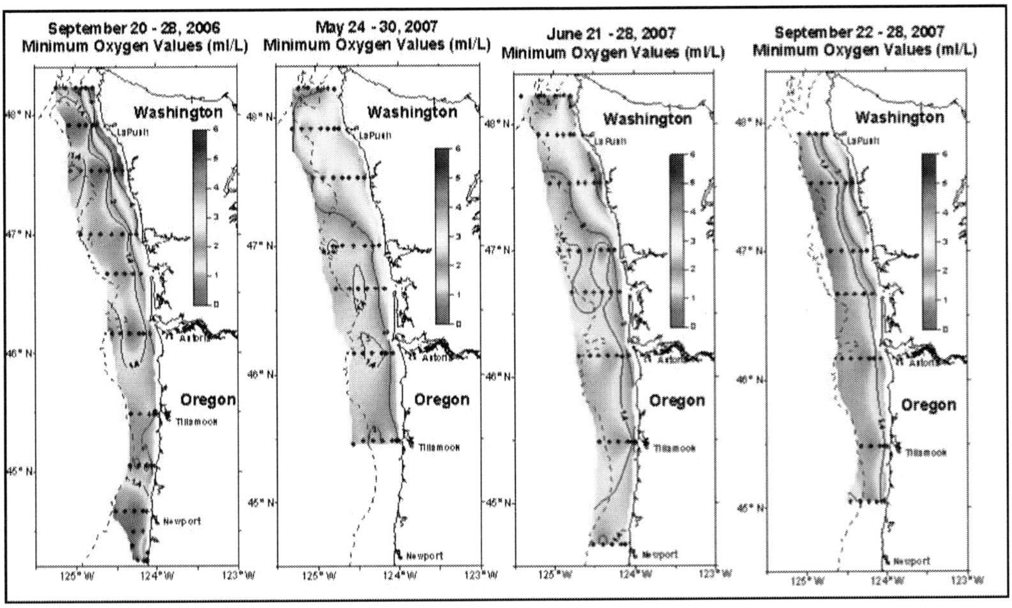

Figure 4. Distribution and extent of hypoxic waters (bold 1.4 ml/L contour) along the Oregon-Washington shelf. Shown is the minimum oxygen value regardless of the depth it occurred (but the minimum values was almost always at the deepest sampling). Panels show the extent in September 2006 (far left), and the seasonal progression during 2007 (May, June and Sept). The dashed line along the coast represents the 200 m isobath. Data Source: Cheryl Morgan (OSU and CIMRS) and Bill Peterson (NOAA NWFSC), project supported by Bonneville Power Administration.

Future Outlook

A coordinated research and monitoring effort aimed at informing resource managers is needed. The combination of increased awareness of the spatial extent of hypoxia, its impact on the shelf ecosystem, and the availability of instruments that can easily be integrated into current scientific programs will provide additional data on the timing and extent of hypoxia along the Oregon-Washington shelf. This improved understanding may also lead to an ability to forecast the extent and severity of low dissolved oxygen along the shelf which could be a useful tool for fishery managers.

Research programs—such as those carried out through the Pacific Coastal Ocean Observing System (NOAA), the Partnership for Interdisciplinary Studies of Coastal Oceans (Oregon State University, et al.), the Oregon Coastal Ocean Observing System (Oregon State University), the Olympic Coast National Marine Sanctuary (NOAA/National Marine Sanctuaries), and the Bonneville Power Administration (Oregon State University, NOAA)—collect hydrographic data, including dissolved oxygen, physical, and biological data along the Oregon-Washington shelf. Combined with regional and broad-scale hydrographic data, it will be possible to gain a better understanding of the potential mechanisms driving the variability of the hypoxic zone.

Understanding how natural versus anthropogenically induced changes in climate influence the spatial extent and severity of hypoxic zones along the shelf requires consistent, long-term observations across the entire region of interest.

REFERENCES

Chan F, Barth JA, Lubchenco J, Kirincich A, Weeks H, Peterson WT, Menge BA. 2008. Emergence of Anoxia in the California Current Large Marine Ecosystem. *Science* 319:920.

Grantham BA, Chan F, Nielsen KJ, Fox DS, Barth JA, Huyer A, Lubchenco J, Menge BA. 2004. Upwelling-driven nearshore hypoxia signals ecosystem and oceanographic changes in the northeast Pacific. *Nature* 429:749-754.

Hickey BM, Banas NS. 2003. Oceanography of the US Pacific Northwest coastal ocean and estuaries with application to coastal ecology. *Estuaries* 26:1010-1031.

Marcus NH. 2001. Zooplankton: Responses to the consequences of hypoxia. In: Rabalais NN, Turner RE (eds) Coastal and Estuarine Studies: Coastal Hypoxia, Vol 58. American Geophysical Union, Washington D.C., p 49-60.

Peterson WT. Unpublished data. National Oceanic and Atmospheric Administration, Northwest Fisheries Science Center.

Sedlacek C, Marcus NH. 2005. Egg production of the copepod *Acartia tonsa*: The influence of hypoxia and food concentration. *J. Exp. Mar. Biol. Ecol.* 318:183-190.

YAQUINA BAY

Physical Description of the System

Yaquina Estuary is a small, drowned, river valley estuary located along the central Oregon coast of the United States (Figure 1) with an estuarine surface area of 19 square kilometers (km^2) and a watershed area of 650 km^2 (Lee et al. 2006). Approximately 48% of the estuarine area is intertidal. November through April (wet season) is a period of high precipitation along the Oregon coast when the estuary is river dominated. From May through October (dry season) the estuary switches from riverine to marine dominance and a saltwedge extends fairly far upriver.

Two tributaries, the Yaquina River and Big Elk Creek, with similarly sized drainage areas, contribute approximately equally to the long-term median freshwater inflow of 7.5 cubic meters per second (m^3/s).

The estuary is well mixed under low flow conditions, and partially- to well-mixed during winter high inflow conditions. Tides are semidiurnal with a mean tidal range of approximately 1.9 m (Shirzad et al. 1989). The estuary is divided into two zones, one of which is dominated by ocean input (Zone 1) and the other which is more influenced by watershed and point source inputs (Zone 2).

Due to the small volume of the estuary and the strong tidal forcing, there is close coupling between the estuary and the coastal ocean. Approximately 70% of the volume of the estuary is exchanged with the coastal ocean during each tidal cycle (from Brown et al. 2007, Bricker et al. 1998). Like other estuaries in the Pacific Northwest that are adjacent to the California Current System, Yaquina Estuary is influenced by seasonal wind-driven upwelling which moves relatively cool, nutrient-rich water into the estuary during April through September.

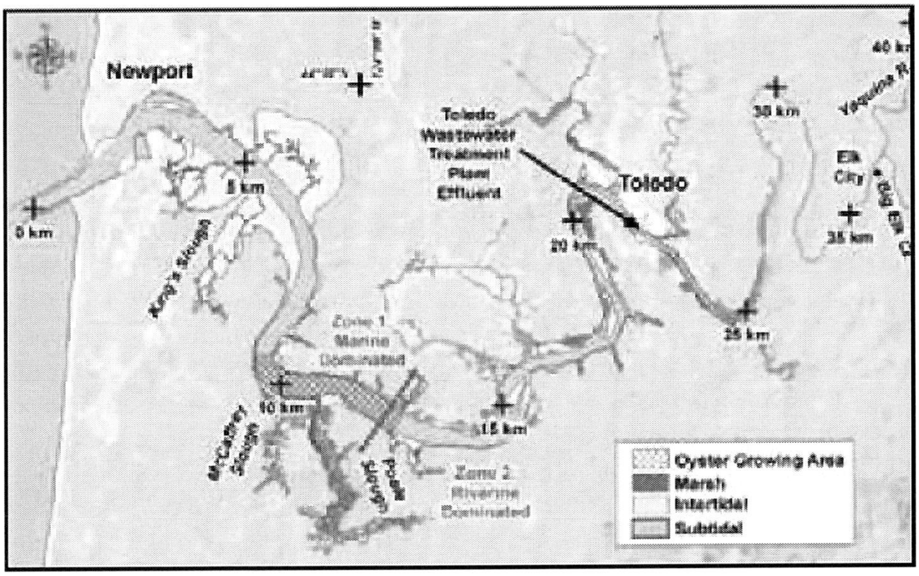

Figure 1. Location map of Yaquina Estuary in Oregon. The estuary is divided into "marine dominated" (Zone 1) and "riverine dominated" (Zone 2) segments (Lee et al. 2006) based on the relative proportion of oceanic-derived nutrients versus terrestrially derived nutrients (from Brown et al. 2007).

History of Hypoxia (Issue, Causes, Economic and Ecosystem Impacts)

The Yaquina watershed is heavily forested with deciduous, evergreen, and shrub land-use classes constituting 85% of the watershed (Lee et al. 2006, based on NOAA 2001 Coastal Change Analysis Program data). Although primarily forested and showing little "urban footprint," the watershed has been impacted by a variety of disturbances, including logging which began in the mid-1800s and continues to the present. In addition to the direct effects of logging on erosion and water quality, rafting of logs can also affect aquatic habitats by physical disturbance, alteration of flow regimes, and accumulation of wood and bark debris and sawdust which can smother the benthos and result in low dissolved oxygen and/or elevated hydrogen sulfide (Sedell et al. 1991). There are three other nutrient sources influencing low dissolved oxygen in the Yaquina Estuary: sewage from municipal discharges, industrial discharges, and nonpoint sewage inputs specifically from septic systems. The Yaquina Estuary watershed contains the cities of Toledo and Newport; however, only the "Bay Front" of Newport lies within the watershed boundaries.

Although the *National Estuarine Eutrophication Assessment* (Bricker et al. 2007) reported that Yaquina has only low level problems with dissolved oxygen, there are strong seasonal patterns within the estuary (Figure 2). Oxygen levels in the estuary are comparatively stable during the wet season, but show a decline during the dry season. The wet season dissolved oxygen values have an overall mean of 9.7 mg/L, which is well above the threshold considered to be hypoxic, indicating no significant problems. The dry season data show a decline to an overall mean value of 5.8 mg/L but it increases again to wet season values. Zone 1 (marine-dominated) and Zone 2 (riverine-dominated) follow the same pattern, but Zone 2 appears to have the lower values overall. The lowest values during the dry season are approaching hypoxic conditions (< 2 mg/L). However, even during the dry season, most values are above the threshold considered to be hypoxic, indicating minimal problems with dissolved oxygen in this system.

During 1960–1984, there was a noted improvement in dissolved oxygen concentrations in Zone 2 (riverine-dominated), but it is not clear if the cause of the observed trend was from the decrease in logging or the improved sewage treatment. Recent (2002-2006) dissolved oxygen levels in Zone 2 are similar to dissolved oxygen levels during the mid-1980s, suggesting that there have been no recent changes in dissolved oxygen levels.

Since 2002, there has been an increase in the incidence of hypoxic events on the Oregon shelf (Grantham et al. 2004) that have the potential to influence dissolved oxygen levels within the estuary, especially during periods of low oxygen water upwelling (and particularly in Zone 1). Data collected 3.7 km from the mouth of the estuary show that hypoxic shelf water is imported into Yaquina Estuary during flood tides. A time series of dissolved oxygen and salinity data measured during July 9-19, 2002, coinciding with a documented hypoxic event on the Oregon shelf off of Newport, Oregon (Grantham et al. 2004), shows the import of hypoxic shelf water into the estuary (Figure 3a). Minimum dissolved oxygen levels occurred during maximum salinities, coincident with flooding tides. In addition, minimum dissolved oxygen levels occur during minimum water temperatures (~ 9 degrees C), which is indicative of recently upwelled water.

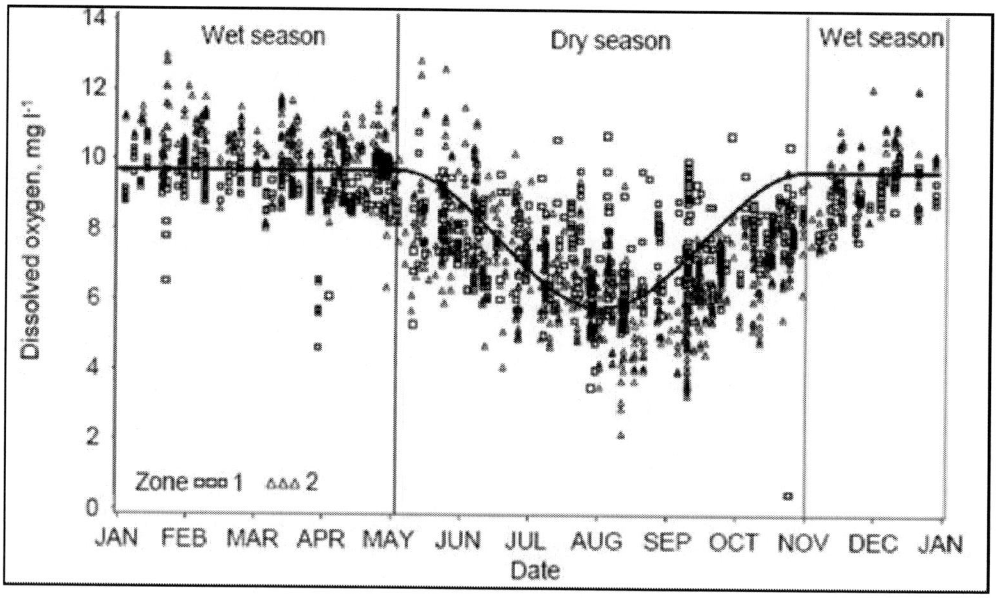

Figure 2. Seasonal pattern (1960-1984) of dissolved oxygen at all locations in the Yaquina Estuary and River with squares and triangles representing samples from Zones 1 (oceanic dominated) and 2 (river dominated), respectively. Solid line is nonlinear least-squares fit to data, which was modeled as a constant during wet season and a cosine function of date during the dry season (n = 869; from Brown et al. 2007).

These results show that, like other Pacific Coast estuaries, dissolved oxygen conditions in the lower portion of the estuary are strongly influenced by ocean conditions due to close coupling between the shelf and the estuary resulting from strong tidal forcing and upwelling during the wet season. Although this has been reported with increasing frequency recently, it is not a new phenomenon. A study of the Yaquina Estuary attributed low dissolved oxygen concentrations (5 mg/L) in the lower estuary to coastal upwelling during July 1968 (Gibson 1974).

Science and Management Actions (to Date and Planned)

In response to observed problems with sewage in the early 1900s, a combined stormwater/sewage system that discharged raw sewage into the Yaquina River was constructed in Toledo in 1926, and then upgraded in 1954 to a primary treatment facility to handle the municipal waste from the city of Toledo. This facility, which discharges into the Yaquina Estuary (about 22 km from the mouth of the estuary), was upgraded to secondary treatment in 1981. In the late 1980s and early 1990s, the city of Toledo made improvements to their stormwater collection system, reducing the bypassing of the treatment plant during high flow periods. In 1996, the Toledo plant had a discharge of 0.979 million gallons per day with a design capacity of 3.5 million gallons per day (www.epa.gov/OW-OWM.html/mtb/cwns/1996report2/ or.htm).

In addition to the Toledo municipal discharge, a number of houses along the Yaquina Estuary and River have on-site septic systems, some which were previously failing. The primary environmental impact of these septic systems appears to be microbial contamination and they have all been repaired. A combined sewage discharge with a pump station was constructed for Newport in the mid-1950s, which eliminated the direct discharge of sewage from Newport into Yaquina Estuary (Brown et al. 2007). A municipal sewage system with primary treatment and an offshore discharge was constructed in Newport in 1964, which has since been upgraded to secondary treatment.

Addressing issues associated with ocean input of nutrients and the seasonal shift in dominance of riverine and oceanic loading is critical in the process of developing nutrient criteria for estuaries in the Pacific Northwest region. Several studies have been conducted to try to understand sources and variability in order to develop appropriate nutrient criteria and management measures.

Future Outlook

Although presently hypoxia is not a significant problem in this estuary, it may worsen in the future since the population of Lincoln County, where Toledo and Newport are located, is predicted to increase by 12% by 2020. It is possible that conditions will remain the same, if wastewater treatment and other land-based nutrient management measures are maintained and improved. However, hypoxia and other nutrient-related problems may worsen due to the additional nutrient loads associated with the population increase if nutrient management infrastructures are overloaded. This is particularly true in Zone 2 (riverine-dominated), since this zone is more influenced by land-based sources of nutrients than Zone 1. Historical reduction of dissolved oxygen in Zone 2 suggests that this system can be impacted by watershed activities even in the presence of strong flushing. However, if the population of Newport increases, water quality in the lower estuary may also worsen as a result. The import of low dissolved oxygen water into the estuary from the coastal ocean may result in this system being susceptible to hypoxia in the future.

REFERENCES

Bricker S, Longstaff B, Dennison W, Jones A, Boicourt K, Wicks C, Woerner J. 2007. Effects of Nutrient Enrichment in the Nation's Estuaries: A Decade of Change, National Estuarine Eutrophication Assessment Update. NOAA Coastal Ocean Program Decision Analysis Series No. 26. National Centers for Coastal Ocean Science, Silver Spring, MD. 322 pp. http://ccma.nos.noaa.gov/news/feature/Eutroupdate.html.

Bricker S, Clement C, Frew S, Harris M, Pirhalla D. 1998. NOAA's Estuarine Eutrophication Survey. Volume 5: Pacific Coast Region. Silver Spring, MD. Office of Ocean Resources Conservation Assessment. 75 pp.

Brown CA, Nelson WG, Boese BL, DeWitt TH, Eldridge PM, Kaldy JE, Lee II H, Power JH, Young DR. 2007. An Approach to Developing Nutrient Criteria for Pacific Northwest Estuaries: A Case Study of Yaquina Estuary, Oregon. USEPA Office of Research and Development, National Health and Environmental Effects Laboratory, Western Ecology Division. *EPA*/600/R-07/046.

Gibson GC. 1974. Oyster mortality study- summary report 1966-1972. Fish Commission of Oregon, Management and Research Division. United States Department of Commerce, *National Marine Fisheries Service*. 37 pp.

Grantham BA, Chan F, Nielsen KJ, Fox DS, Barth JA, Huyer A, Lubchenco J, Menge BA. 2004. Upwelling-driven nearshore hypoxia signals ecosystem and oceanographic changes in the northeast Pacific. *Nature* 429:749-754.

Lee H. II, Brown CA, Boese BL, Young DR (eds.). 2006. Proposed Classification Scheme for Coastal Receiving Waters Based on SAV and Food Web Sensitivity to Nutrients, Volume 2: Nutrient Drivers, Seagrass Distributions, and Regional Classifications of Pacific Northwest Estuaries, United States Environmental Protection Agency Report, Office of Research and Development, National Health and Environmental Effects Laboratory. Internal Report.

Sedell JR, Leone N, Duval WS. 1991. Water Transportation and Storage of Logs. *In:* Influences of Forest and Rangeland Management on Salmonid Fishes and Their Habitats. *American Fisheries Society Special Publication* 19:325-368.

Shirzad FF, Orlando SP, Klein CJ, Holliday SE, Warren MA, Monaco ME. 1989. National estuarine inventory: Supplement 1, Physical and Hydrologic characteristics, The Oregon estuaries., National Oceanic and Atmospheric Administration, Rockville, MD.

HOOD CANAL

Physical Description of the System

Despite its name, Hood Canal is a natural formation which has experienced hypoxia periodically dating back centuries (Brandenberger et al. 2009). It is a sub-basin of Puget

Sound, Washington, with a fjord-like structure (Figure 1), including a natural sill at the mouth which restricts circulation with greater Puget Sound; it has a surface area of 386 square kilometers (km^2) (King County 2001). Restricted and slow circulation coupled with a strong salinity stratification and high productivity make it conducive to low dissolved oxygen conditions (Newton 2007).

History of Hypoxia (Issue, Causes, Economic and Ecosystem Impacts)

Hypoxic events have increased in intensity, duration, and spatial extent since the 1990s, causing fish kills that prompted the Washington State Department of Fish and Wildlife to close many fisheries in Hood Canal in 2003. Southern Hood Canal, towards Lynch Cove (Figure 2), has experienced more hypoxia than the rest of the canal, and the measured dissolved oxygen concentrations are now lower in this area than they were in the 1950s and 1960s (Newton 2007). Hypoxia also develops seasonally along the main stem of Hood Canal, and transient upwelling of these waters caused by wind events has caused fish kills that garnered considerable public attention. Mortality events of living resources have been reported back to the 1920s (Fagergren et al. 2004). Ecosystem impacts such as these, however, have increased in frequency since 2002 including an extensive event in September 2006 (Newton 2007).

Nitrogen inputs, especially from the Lynch Cove area, have been implicated, but nitrate from inflowing seawater (Paulson et al. 2006), changes in circulation and flushing (e.g., modified river flows), stratification, and algal blooms are also important (HCDOP 2007).

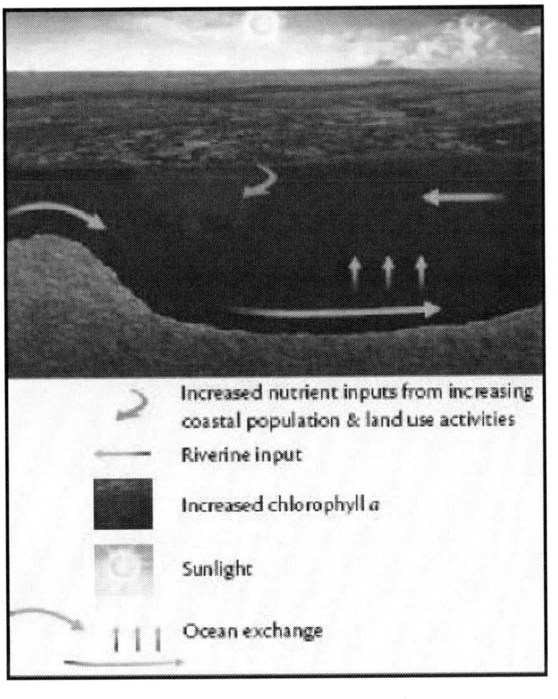

Figure 1. Diagram of Cross Section of Hood Canal with prevailing geography and physical conditions (Newton 2007).

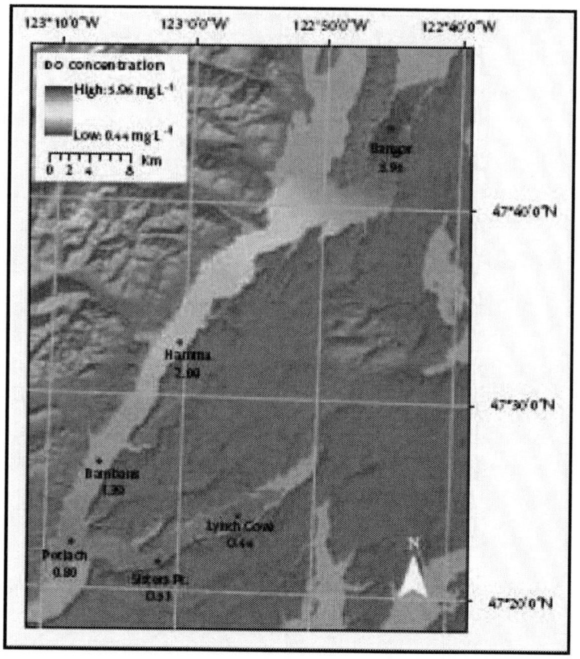

Figure 2. August 2006 interpolation reflecting typical pattern of low DO concentrations in Hood Canal (Newton 2007).

Science and Management Actions (to Date and Planned)

As a result of heightened attention to Hood Canal environmental deterioration, two significant developments may lead to improvements in hypoxia in the future. First, the Hood Canal Dissolved Oxygen Program (http://www. hoodcanal.washington. edu/index.jsp), a partnership of 28 organizations, including NOAA, EPA, USGS, USACE, and the US Navy, was created in 2005 to study oxygen dynamics through broad-scale community involvement, concerted assessments, and development of models. In addition to the Hood Canal Dissolved Oxygen Program, but operating on a larger scale, the Puget Sound Partnership (http://www.psp.wa.gov/), a new state agency, was established in 2007 to develop a coordinated, region-wide response to the deterioration of the Sound. The group adopted an ambitious action plan in 2008.

Future Outlook

The factors driving hypoxia in Hood Canal are still being clarified, especially the role of nutrients derived from human activities. Creation of the Puget Sound Partnership in 2007—which represents a broad coalition of citizen, Federal, state, and tribal agencies and interests—should allow development of deliberate protection actions for Puget Sound, and Hood Canal specifically.It is important to develop a comprehensive management plan before the situation worsens given the expected growth in the coastal population around Puget Sound.

REFERENCES

Brandenberger JM, Louchouarn P, Crecelius EA. 2009, *submitted*. Historical reconstruction of redox sensitive metals in two basins of Puget Sound: Evaluating Natural and Post Urbanization Signatures of Hypoxia. Submitting to *Geochimica et Cosmochimica Acta.*

Fagergren D, Criss A, Christensen D. 2004. Hood Canal low dissolved oxygen: Preliminary assessment and corrective action plan: Olympia, Washington, Puget Sound Action Team, Publication #PSAT04-06, 43 pp. plus appendices.

HCDOP (Hood Canal Dissolved Oxygen Program). 2007. http://www. Hood canal. washington.edu/observations/ historicalcomparison.jsp.

King County. 2001. Features Of Puget Sound Region: Oceanography And Physical Processes, Chapter 3 of the State of the Nearshore Report, King County Department of Natural Resources, Seattle, WA, 2001.

Newton JA. 2007. Hood Canal, WA: The complex factors causing low dissolved oxygen events require ongoing research, monitoring, and modeling. In: Bricker S, Longstaff B, Dennison W, Jones A, Boicourt K, Wicks C, Woerner J. 2007. Effects of Nutrient Enrichment in the Nation's Estuaries: A Decade of Change, National Estuarine Eutrophication Assessment Update. NOAA Coastal Ocean Program Decision Analysis Series No. 26. National Centers for Coastal Ocean Science, Silver Spring, MD. 322 pp. http://ccma.nos.noaa.gov/news/feature/Eutroupdate.html.

Paulson AJ, Konrad CP, Frans LM, Noble M, Kendall C, Josberger EG, Huffman RL, Olsen TD. 2006. Freshwater and saline loads of dissolved inorganic nitrogen to Hood Canal and Lynch Cove, western Washington: US Geological Survey Scientific Investigations Report 2006-5106, 92 pp.

APPENDIX III. TABLE OF US SYSTEMS IMPACTED BY HYPOXIA

Table A1. Systems in the United States by region and their dissolved oxygen condition

Regions are based on Bricker et al. 2007. Analysis current through December 2008. Documented indicates systems with low dissolved oxygen measurements leading to hypoxia and evidence that hypoxia was caused, at least in part, by human activity—mainly related to eutrophication, but also hydraulic modifications such as dams and canal construction.

Concern indicates systems that exhibit primary or secondary effects of eutrophication, including elevated nutrient levels, elevated chlorophyll levels, or harmful algal blooms.

These systems are impaired by nutrients and are possibly at risk of developing hypoxia.

Some of the systems classified as areas of concern may already be experiencing hypoxia, but there is a lack dissolved oxygen data.

None indicates systems with no measured low dissolved oxygen and with low expression of primary or secondary effects of eutrophication.

Table A1. Systems in the United States by region and their dissolved oxygen condition

State	System	Condition	Decade Condition First Recorded	Faunal Response	Trends and Other Comments	Reference
			Great Lakes Region			
Michigan	Lake Huron, Central Basin	None	1990			Ahrnsbrak and Wing 1998
Michigan	Lake Michigan, Central Basin	None	1990			Ahrnsbrak and Wing 1998
Michigan	Lake Superior, Central Basin	None	1990			Ahrnsbrak and Wing 1998
New York	Lake Ontario, Kingston Basin	Documented	1990			Ahrnsbrak and Wing 1998
New York	Lake Ontario, Main Basin	None	1990			Ahrnsbrak and Wing 1998
New York	Lake Ontario, Prince Edward Bay	Concern	1990			Ahrnsbrak and Wing 1998
Ohio	Lake Erie, Central Basin	Documented	1700	Suitable whitefish habitat reduced by low dissolved oxygen. Benthos reduced.	Periodic hypolimnetic anoxia prior to European settlement. Global warming will lower dissolved oxygen from higher microbial metabolism. Increase in anoxia	Hartman 1972, Krieger 1985, Rosa and Burns 1987, Delorme 1982, Edwards et al. 2005, Blumberg and DiToro 1990, Krieger et al. 2007
					from 1930s to 1970s.	
Ohio	Lake Erie, Eastern Basin	None	1990			Ahrnsbrak and Wing 1998
Ohio	Lake Erie, Western Basin	Documented	1950	Mortality of mayflies in 1953. Mayfly populations partially recovered in 1990s.	Occasional hypoxia, improved in 1990s.	Britt 1955, Krieger 1985, Gerlofsma and Ciborowski 1999
			Gulf of Mexico Region			

State	System	Condition	Decade Condition First Recorded	Faunal Response	Trends and Other Comments	Reference
Alabama	Alabama Inner Shelf	Documented	1990		Area off Dauphin Island hypoxic. Dissolved oxygen declined from 1980s to 1990s.	Rabalais et al. 1985, Rabalais 1998, Byrnes et al. 1999
Alabama	Bon Secour Bay	Documented	2000	Mortality of benthos and loss of oysters.		Rikard et al. 2000
Alabama	Meaher Park	Documented	2000		Data is from water quality monitoring stations.	Mobile Bay NEP 2008
Alabama	Mobile Bay	Documented	1860	Occasional shoreward migration of hypoxia stressed fish and crustaceans (Jubilees).	May have been hypoxic in 1800s. Has winter hypoxia. Stratification a major factor. Some improvement in dissolved oxygen between 1900s and 2000s.	Pennock et al. 1994, May, 1973, Loesch 1960, Engle and Summers 1999, Rabalais et al. 1985, Bricker et al. 1999, 2007
Alabama	Mobile River	Concern	1980			Rabalais et al. 1985
Alabama	Tensaw River	Concern	1980			Rabalais et al. 1985
Alabama	Weeks Bay	Documented	1990		Mangrove area.	Sanger et al. 2002
Alabama	Wolf Bay	Documented	1980		Part of regional water quality monitoring program.	Engle et al. 1999, Rabalais et al. 1985
Florida	Alafia River	Documented	1980			Windsor 1985
Florida	Alaqua Creek	None	1980			Windsor 1985
Florida	Allen Creek	Concern	1980			Windsor 1985
Florida	Alligator Alley Canal	Documented	1970			Windsor 1985
Florida	Alligator Creek	Concern	1980			Windsor 1985

Table A1. (Continued)

State	System	Condition	Decade Condition First Recorded	Faunal Response	Trends and Other Comments	Reference
Florida	Apalachee Bay	Documented	1970		Localized hypoxia and anoxia in 1970s attributed to pulp mills in Econfina and	Windsor 1985, Bricker et al. 1999, Heck 1976, Livingston 1975
					Fenholloway Rivers, some dissolved oxygen improvements from1990s to 2000s.	
Florida	Apalachicola Bay	Documented	1990		No hypoxia in 1980s, observed in 1990s. Some improvement in dissolved oxygen in 2000s.	Lowery 1998, Sanger et al. 2002, Windsor 1985, Bricker et al. 1999
Florida	Apalachicola River	None	1980			Windsor 1985
Florida	Aucilla River	Concern	1980			Windsor 1985
Florida	Bayou Chico	Documented	1990			Summers et al. 1997
Florida	Bayou Marcus Creek	Documented	1980			Windsor 1985
Florida	Bear Creek	Concern	1980			Windsor 1985
Florida	Big Lagoon	Documented	1990		Not hypoxic in 1980s. Part of regional water quality monitoring program.	Whitledge 1985, Engle et al. 1999
Florida	Big Lostman's Bay	Concern	1980			Windsor 1985
Florida	Blackwater River	None	1980			Windsor 1985
Florida	Caloosahatchee River	Documented	1990			Windsor 1985, Bricker et al. 1997, 2007
Florida	Carpenters Creek/Bayou Texar	Concern	1980			Windsor 1985
Florida	Carrabelle River	None	1980			Windsor 1985

State	System	Condition	Decade Condition First Recorded	Faunal Response	Trends and Other Comments	Reference
Florida	Charlotte Harbor	Documented	1980	1980s to 1990s improvement in water quality from nutrient management, 2000s decline in water quality as population expanded.	Dissolved oxygen declined from 1970s to 1990s, than increased to 1970s levels by end of 1990s, declined again in 2000s. Associated wet season and stratification, hurricane disturbance resulted in more low dissolved oxygen.	Tomasko et al 2006, Turner et al. 2006, Windsor 1985
Florida	Chatham Bay/ River	Concern	1980			Windsor 1985
Florida	Choctawhatchee Bay	Documented	1980		Some improvement in dissolved oxygen between 1990s and 2000s.	Lowery 1998, Windsor 1985, Bricker et al. 1999, 2007
Florida	Choctawhatchee River	None	1980			Windsor 1985
Florida	Clowers Creek	Documented	1980			Windsor 1985
Florida	Cross Bayou Canal	Documented	1980			Windsor 1985
Florida	Crystal Bay	Concern	1980			Windsor 1985
Florida	Crystal River	Concern	1980			Windsor 1985
Florida	Deadman Bay	None	1980			Windsor 1985
Florida	Deerpoint Lake	Concern	1980			Windsor 1985
Florida	Delaney Creek	Documented	1990			Windsor 1985
Florida	East Bay, Pensacola	Documented	1980		Some improvement in dissolved oxygen between 1980s and 2000s. Improved from nutrient management, municipal and industrial.	Windsor 1985

Table A1. (Continued)

State	System	Condition	Decade Condition First Recorded	Faunal Response	Trends and Other Comments	Reference
Florida	East Bay, Apalachicola	None	1980			Windsor 1985
Florida	Econfina Creek	None	1980			Windsor 1985
Florida	Econfina River	Documented	1970		Some hypoxia in 1970s, dissolved oxygen improved with effluent control. Winter and summer hypoxia.	Windsor 1985, Livingston 1975
Florida	Eleven Mile Creek	Documented	1980			Windsor 1985
Florida	Escambia Bay	Documented	1960	Loss of fisheries species, fish kills.		Summers et al. 1997, Windsor 1985
Florida	Escambia River	None	1980			Windsor 1985
Florida	Everglades GAC Canal	Documented	1970			Windsor 1985
Florida	Fenholloway River	Documented	1970		Hypoxia and anoxia in 1970 attributed to pulp mill effluents. Dissolved oxygen improved with effluent control. Winter and summer hypoxia and anoxia.	Windsor 1985, Livingston 1975
Florida	Florida Bay	Documented	2000		Some improvement in dissolved oxygen between 1990s and 2000s but hypoxic events may be more frequent in 2000s.	Bricker et al. 1999, 2007, TIEE 2008
Florida	Gordon River	Documented	1980			Windsor 1985
Florida	Hillsborough Bay	Documented	1970	Almost complete	Improved from	Santos and Simon 1980,

State	System	Condition	Decade Condition First Recorded	Faunal Response	Trends and Other Comments	Reference
				mortality of benthos in late summer.	nutrient management.	Windsor 1985, Bricker et al. 2007
Florida	Hillsborough River	Documented	1980			Windsor 1985
Florida	Inner Continental Shelf	Documented	2000	Mortality of benthos. Mortality of fishes from combined effects of HAB and low dissolved oxygen.	Low dissolved oxygen from HAB (toxic dinoflagellate Karenia brevis) in August 2005. Multiple stressors involved in mortalities.	Landsberg 2009
Florida	Jones and Jackson Creeks	Documented	1980			Windsor 1985
Florida	Lake Panasoffkee	None	1980			Windsor 1985
Florida	Lake Rousseau	None	1980			Windsor 1985
Florida	Little Aucilla River	Documented	1980			Windsor 1985
Florida	Long Bayou	Documented	1980			Windsor 1985
Florida	Looe Key	Documented	1980	Mortality of benthos and corals.	General eutrophication affecting water quality, dissolved oxygen conditions worsened from 1990s to 2000s. Rainfall events followed by low dissolved oxygen in seagrass and coral reef areas.	Lapointe and Matzie 1996, Whitledge 1985, Bricker et al. 2007
Florida	Martin Bayou	Concern	1980			Windsor 1985
Florida	Mulatto and Indian Bayous	Documented	1980			Windsor 1985
Florida	Munson Slough	Documented	1980			Windsor 1985

Table A1. (Continued)

State	System	Condition	Decade Condition First Recorded	Faunal Response	Trends and Other Comments	Reference
Florida	Myakka River	Documented	1980		Combination of agricultural runoff and swamp drainage.	Windsor 1985
Florida	North Bay	Concern	1980			Windsor 1985
Florida	Ochlockonee Bay	None	1980			Windsor 1985
Florida	Ochlockonee River	None	1980			Windsor 1985
Florida	Old River	Documented	1990		Part of regional water quality monitoring program.	Engle et al. 1999
Florida	Otter Creek	None	1980			Windsor 1985
Florida	Patch Reef	Documented	1990	Mortality of benthos.	Rainfall events followed by low dissolved oxygen in seagrass and coral reef areas.	Lapointe and Matzie 1996
Florida	Peace River	Documented	1980	Hurricanes worsened hypoxia in 2004.	Worse after hurricanes in 2004. Algal blooms from nutrients and sewage.	Windsor 1985, Tomasko et al. 2006, Turner et al. 2006
Florida	Pensacola Bay	Documented	1960	Loss of fisheries species, fish kills. Reduction in shrimp catch starting in 1960s attributed to multiple stressors including low dissolved oxygen.	Some improvement in dissolved oxygen between 1980s and 2000s.	Lowery 1998, Hagy et al. 2006, Windsor 1985, Hagy and Murrell 2007
Florida	Perdido Bay	Documented	1960		Dissolved oxygen poor in 1960s, but	Flemer et al. 1999, Whitledge 1985, Bricker

State	System	Condition	Decade Condition First Recorded	Faunal Response	Trends and Other Comments	Reference
					improved 1970s to 1980s. Dissolved oxygen declined from 1990s to 2000s. Stratification and industrial effluents highly significant factors, high nutrient loads.	et al. 1999, 2007
Florida	Perdido River	Documented	1960			Windsor 1985
Florida	Philippi Creek	Documented	1980			Windsor 1985
Florida	Pine Channel	Documented	1990	Mortality of benthos.	Rainfall events followed by low dissolved oxygen in seagrass and coral reef areas.	Lapointe and Matzie 1996
Florida	Port Pine	Documented	1990	Mortality of benthos.	Rainfall events followed by low dissolved oxygen in seagrass and coral reef areas.	Lapointe and Matzie 1996
Florida	Rocky Creek	None	1980			Windsor 1985
Florida	Rookery Bay	Documented	1990		Dissolved oxygen levels declined from 1990s to 2000s.	Rabalais 1998, Sanger et al. 2002, Bricker et al. 1999, 2007
Florida	Santa Rosa Sound	None	1980			Windsor 1985
Florida	Sarasota Bay	Documented	1980	SAV recovering.	Dissolved oxygen levels improved from 1990s to 2000s.	Windsor 1985, Bricker et al. 1999, 2007
Florida	Sopchoppy River	None	1980			Windsor 1985
Florida	Spring Warrior Creek	None	1980			Windsor 1985

Table A1. (Continued)

State	System	Condition	Decade Condition First Recorded	Faunal Response	Trends and Other Comments	Reference
Florida	St. Andrew Bay	Documented	1970		Hypoxia reduced from nutrient management, municipal and industrial.	Windsor 1985
Florida	St. George Sound	Documented	1990	No dissolved oxygen problem in 1980s.	Part of regional water quality monitoring program.	Engle et al. 1999, Windsor 1985
Florida	St. Joseph Bay	Documented	1990	Avoidance by mobile fauna and brittlestar migration.	Was not hypoxic in 1980s.	Windsor 1985, Leonard and McClintock 1999
Florida	St. Marks River	None	1980			Windsor 1985
Florida	Steinhatchee River	None	1980			Windsor 1985
Florida	Suwannee River	Concern	1980			Windsor 1985
Florida	Tampa Bay	Documented	1980	SAV recovering.	Improved conditions for dissolved oxygen from 1980s to 1990s, dissolved oxygen worsened from 1990s to 2000s. Improved from nutrient management.	Leverone 1995, Lowery 1998, Bricker et al. 1999, 2007
Florida	Ten Mile Creek	None	1980			Windsor 1985
Florida	Ten Thousand Islands North	Documented	2000			Bricker et al. 1999, 2007
Florida	Ten Thousand Islands South	Documented	2000			Bricker et al. 1999, 2007
Florida	Turkey Creek	None	1980			Windsor 1985
Florida	Useppa Island	Documented	1940	Some fish mortality.	Low dissolved oxygen from HAB	Gunter et al. 1948

State	System	Condition	Decade Condition First Recorded	Faunal Response	Trends and Other Comments	Reference
					(Gymnodinium).	
Florida	Waccasassa Bay	None	1980			Windsor 1985
Florida	Wakulla River	Concern	1980			Windsor 1985
Florida	Watson Bayou	Documented	1980		Part of regional water quality monitoring program.	Windsor 1985, Engle et al. 1999
Florida	Weeki Wachee Springs	Documented	1980			Windsor 1985
Florida	West Bay	Concern	1980			Windsor 1985
Florida	Whitaker Bayou	Documented	1980			Windsor 1985
Florida	Withlacoochee River	Concern	1980			Windsor 1985
Florida	Yellow River	None	1980			Windsor 1985
Louisiana	Amite River	Documented	1990	Reduced benthos and loss of large clams.		Penland et al. 2002 (http://pubs.usgs.gov/of/2002/of02-206/index.html)
Louisiana	Atchafalaya Bay	Documented	2000		Not hypoxic in 1980s. Stratification highly significant factor.	Rabalais et al. 1985, Bricker et al. 2007
Louisiana	Atchafalaya River	None	1980			Rabalais et al. 1985, Bricker et al. 2007
Louisiana	Balize Delta	Documented	1950		Low dissolved oxygen present in 1950s based on foraminifera proxies.	Brunner et al. 2006
Louisiana	Barataria Bay	Documented	1950		Not hypoxic in 1980s.	Rabalais et al. 1985, Bricker et al. 2007, Engle et al. 1999
Louisiana	Blind Bay	Concern	1990			Engle et al. 1999
Louisiana	Breton Sound	Documented	1990		Not hypoxic in 1980s.	Rabalais et al. 1985, Bricker et al. 2007, Engle et al. 1999

Table A1. (Continued)

State	System	Condition	Decade Condition First Recorded	Faunal Response	Trends and Other Comments	Reference
Louisiana	Caillou Bay	None	1980			Rabalais et al. 1985
Louisiana	Calcasieu Estuary	Concern	1980			Whitledge 1985
Louisiana	Calcasieu Lake	Concern	1980		Some improvement in dissolved oxygen levels between 1990s and 2000s.	Rabalais et al. 1985, Bricker et al. 1999, 2007
Louisiana	Calcasieu Pass/ Channel	Concern	1980			Rabalais et al. 1985, Bricker et al. 2007
Louisiana	Chandeleur Sound	Documented	1990		Not hypoxic in 1980s.	Rabalais et al. 1985, Bricker et al. 2007, Engle et al. 1999
Louisiana	East Cote Blanche Bay	None	1980			Rabalais et al. 1985
Louisiana	Graden Island Bay	Documented	1990		Part of regional water quality monitoring program.	Engle et al. 1999
Louisiana	Grand Bay	Documented	1990		Part of regional water quality monitoring program.	Engle et al. 1999
Louisiana	Lake Boudreaux	None	1980			Rabalais 1998
Louisiana	Lake Cataouatche	Documented	1970	Fish kills.		Rabalais 1998
Louisiana	Lake Maurepas	None	1970			Rabalais 1998
Louisiana	Lake Penchant	Concern	1980	Low dissolved oxygen may have caused bass to leave the lake.		Rabalais et al. 1985
Louisiana	Lake Pontchartrain	Documented	1970	Reduced species diversity. Loss of large clams.	250 km2 USGS estimate.	Abadie and Poirrier 2000, Rabalais et al. 1985, Penland et al. 2002
Louisiana	Lake Tambour	None	1980			Rabalais 1998
Louisiana	Mermentau River	None	1990			Bricker et al. 1997

State	System	Condition	Decade Condition First Recorded	Faunal Response	Trends and Other Comments	Reference
Louisiana	Mississippi/ Atchafalaya River Plumes	Documented	1970	Avoidance by mobile fauna. Mortality of benthos with annual recolonization, Loss of biomass.	Area coincides with historic white and brown shrimp fishing grounds. Annual hypoxia developed in 1970s, is now the largest hypoxic area in US. Major events in Mississippi watershed may have led to low dissolved oxygen prior to 1700s.	Turner and Rabalais 1994, Justic et al. 1996, Sen Gupta et al. 1996, Rabalais et al. 2007, Rabalais and Turner 2001, Osterman et al. 2007, Craig and Crowder 2005
Louisiana	Moss Bay	None	1980			Rabalais 1998
Louisiana	Terrebonne Bay	Concern	1980			Rabalais et al. 1985
Louisiana	Timbalier Bay	None	1980			Rabalais et al. 1985
Louisiana	Vermilion Bay	Documented	2000		Not hypoxic in 1980s. Stratification highly significant factor.	Rabalais et al. 1985, Bricker et al. 2007
Louisiana	Wax Lake	Documented	1990		Part of regional water quality monitoring program.	Engle et al. 1999
Louisiana	West Cote Blanche Bay	None	1980			Rabalais et al. 1985
Louisiana	West Hackberry	Documented	1980	Mortality with annual recolonization.	Associated with brine seep.	Gaston 1985
Mississippi	Back Bay	Documented	1970		Low dissolved oxygen occurs but not a recurring problem in 1980s. Part of regional water quality monitoring program.	Engle et al. 1999, Rabalais et al. 1985
Mississippi	Bayou Casotte/ Chico	Documented	1970			Rabalais 1998
Mississippi	Bayou La Batre	Concern	1980			Rabalais 1998

Table A1. (Continued)

State	System	Condition	Decade Condition First Recorded	Faunal Response	Trends and Other Comments	Reference
Mississippi	Big Lake	Documented	1980		Episodic algal blooms.	Rabalais et al. 1985, Rabalais 1998
Mississippi	Biloxi Bay	Concern	1970		Low dissolved oxygen occurs but not a recurring problem in 1980s.	Engle et al. 1999, Rabalais et al. 1985
Mississippi	Biloxi River	Concern	1990			Rabalais 1998
Mississippi	Escatawpa River	Documented	1970			Rabalais et al. 1985
Mississippi	Jourdan River	Concern	1990			Rabalais 1998
Mississippi	Lake Borgne	Concern	1980			Rabalais et al. 1985
Mississippi	Little River	Concern	1980			Rabalais 1998
Mississippi	Mississippi Inner Shelf, Deeper	None	1950		Based on foraminifera proxies, there was no low dissolved oxygen in deeper areas >50 m.	Brunner et al. 2006
Mississippi	Mississippi Inner Shelf, Shallow	Documented	1950		Sporadic low dissolved oxygen in 1800s associated with flood events, by 1940s low dissolved oxygen increased and intensified in 1960s based on foraminifera proxies.	Rabalais et al. 1985, Brunner et al. 2006
Mississippi	Mississippi River, Lower	None	1980			Rabalais et al. 1985
Mississippi	Mississippi Sound East	Documented	1980		Dissolved oxygen improved from 1990s to 2000s. Stratification highly	Rabalais et al. 1985, Bricker et al. 1999, 2007

State	System	Condition	Decade Condition First Recorded	Faunal Response	Trends and Other Comments	Reference
					significant factor.	
Mississippi	Mississippi Sound West	Documented	1950		Low dissolved oxygen in 1950s, intensified in 1960s based on foraminifera proxies.	Rabalais et al. 1985, Bricker et al. 2007, Brunner et al. 2006
Mississippi	Pascagoula Bay	Concern	1980		Dissolved oxygen declined from 1980s to 1990s.	Rabalais et al. 1985, Rabalais 1998
Mississippi	Pascagoula River	Documented	1970		Hypoxic from March to October.	Rabalais 1998
Mississippi	Pearl River	None	1980			Rabalais et al. 1985, Rabalais 1998
Mississippi	St. Louis Bay	None	1980			Rabalais et al. 1985
Mississippi	Tchoutacabouffa River	Concern	1990			Rabalais 1998
Mississippi	West Fowl River	Concern	1980			Rabalais 1998
Mississippi	Wolf River	None	1990			Rabalais 1998
Texas	Alazan Bay	Concern	1980			Rabalais 1998
Texas	Aransas Bay	Documented	2000	Diel hypoxia alters nursery value of SAV beds. Larval red drum, Sciaenops ocellatus, growth reduced when exposed to hypoxia >8 hours.	Open Bay had little low dissolved oxygen in 1980s. Increased nutrient loads lower dissolved oxygen in SAV beds.	Lowery 1998, Rabalais et al. 1985, Perez-Domingues et al. 2006
Texas	Arroyo Colorado	Documented	1980		Hypoxia or anoxia occur year- round.	Engle et al. 1999, Rabalais et al. 1985
Texas	Baffin Bay	Documented	1980			Rabalais et al. 1985, Bricker et al. 2007
Texas	Brazos River	Documented	1980			Rabalais et al. 1985, Bricker et al. 2007
Texas	Brazos Santiago Pass	None	1980			Rabalais 1998

Table A1. (Continued)

State	System	Condition	Decade Condition First Recorded	Faunal Response	Trends and Other Comments	Reference
Texas	Brownsville Ship Channel	Documented	1980			Rabalais 1998
Texas	Bryan Mound, deep	Documented	1980	Decreased diversity and abundance of mobile species at the onset of hypoxia. Mortality of benthos with annual recolonization.		Harper et al. 1981, 1991
Texas	Bryan Mound, shallow	Documented	1970	Stressed fishes, mass mortality of benthos with multi-year recovery.		Harper et al. 1981, 1991
Texas	Cayo del Grullo	Concern	1980			Rabalais 1998
Texas	Cedar Bayou	Documented	1980			Rabalais et al. 1985
Texas	Chocolate Bay	Documented	1990		Part of regional water quality monitoring program.	Engle et al. 1999
Texas	Colorado River	None	1980			Rabalais et al. 1985
Texas	Copano Bay	None	1980			Rabalais 1998
Texas	Corpus Christi Bay	Documented	1980	Reduced, biomass at hypoxic stations is 96% less than at stations with normal oxygen levels.	Hypoxia expanded in the 1990s. Low bottom dissolved oxygen when bottom salinity was high.	Ritter and Montagna 1999, Montagna and Ritter 2006, Montagna and Kalke 1992, Applebaum et al. 2005
Texas	Cox Bay	None	1980			Rabalais 1998
Texas	Dickinson Bayou	Documented	1980			Rabalais et al. 1985
Texas	East Bay	Documented	1980			Whitledge 1985
Texas	Espiritu Santo Bay	None	1980			Rabalais 1998
Texas	Freeport	Documented	1970	Avoidance, some		Harper and Rabalais 1995

State	System	Condition	Decade Condition First Recorded	Faunal Response	Trends and Other Comments	Reference
				mortality of fishes. Mortality of benthos with multi-year recovery.		
Texas	Freeport Ship Channel	None	1980			Rabalais et al. 1985
Texas	Galveston Bay	Concern	1970			Lowery 1998, Rabalais et al. 1985
Texas	Guadalupe Estuary	None	1980		Lowest frequency of low dissolved oxygen of all Texas estuaries in 1980s.	Rabalais et al. 1985, Rabalais 1998
Texas	Keller Bay	None	1980			Rabalais 1998
Texas	Laguna Madre, Lower	Documented	1980		Sewage and shrimp farming, most low dissolved oxygen in deeper areas, diel low dissolved oxygen in shallow areas.	Ziegler and Benner 1998, Rabalais et al. 1985, Bricker et al. 2007
Texas	Laguna Madre, Upper	Documented	1980		Stratification highly significant factor, most low dissolved oxygen in deeper areas.	Rabalais et al. 1985, Bricker et al. 2007
Texas	Laguna Salada	Concern	1980			Rabalais 1998
Texas	Lavaca Bay	Documented	1990		Part of regional water quality monitoring program.	Engle et al. 1999
Texas	Matagorda Bay	Documented	1980			Lowery 1998, Rabalais et al. 1985, Bricker et al. 2007
Texas	Matagorda Ship Channel	Documented	1980			Rabalais 1998
Texas	Mesquite Bay	None	1980			Rabalais 1998

Table A1. (Continued)

State	System	Condition	Decade Condition First Recorded	Faunal Response	Trends and Other Comments	Reference
Texas	Mission Bay	None	1980			Rabalais 1998
Texas	Mission-Aransas Estuary	Documented	1980			Rabalais et al. 1985, Rabalais 1998
Texas	Nueces Bay	Documented	1980			Rabalais 1998
Texas	Nueces Estuary	Concern	1980			Rabalais 1998
Texas	Offatts Bayou	Documented	1930	Mass mortality of benthos and fishes.	First occurrence 1936. Part of regional water quality monitoring program.	Gunter 1942, Engle et al. 1999, White et al. 1984
Texas	Oso Bay	Concern	1980			Rabalais 1998
Texas	Port Arthur Canal	Concern	1980			Rabalais et al. 1985
Texas	Redfish Bay	None	1980			Rabalais 1998
Texas	Rio Grand Tidal	Concern	1980			Rabalais et al. 1985
Texas	Sabine Lake	Concern	1980			Rabalais et al. 1985
Texas	Sabine Pass	Concern	1980			Rabalais et al. 1985
Texas	Sabine-Neches Estuary	Documented	1980			Rabalais et al. 1985
Texas	San Antonio Bay	Documented	1980		Dissolved oxygen declined from 1980s to 1990s. Stratification highly significant factor.	Lowery 1998, Rabalais et al. 1985
Texas	San Antonio River	None	1980			Rabalais 1998
Texas	San Bernard River	Documented	1990		Part of regional water quality monitoring program.	Engle et al. 1999
Texas	South Bay	None	1980			Rabalais 1998
Texas	Texas Inner Shelf	Concern	1980			Rabalais et al. 1985
Texas	Tres Palacios Bay	None	1980			Rabalais 1998
Texas	Trinity Bay	Concern	1980			Whitledge 1985

State	System	Condition	Decade Condition First Recorded	Faunal Response	Trends and Other Comments	Reference
Texas	Trinity-San Jacinto Estuary	Concern	1990			Rabalais 1998
Texas	Upper Galveston Bay Channel	Documented	1980		Part of regional water quality monitoring program.	Engle et al. 1999, Rabalais et al. 1985
Texas	West Bay	Concern	1980			Rabalais et al. 1985
	Mid-Atlantic Region					
Connecticut	Ash Creek	Documented	1980			Whitledge 1985
Connecticut	Black Rock Harbor	None	1980			Whitledge 1985
Connecticut	Bridgeport Harbor	None	1980			Whitledge 1985
Connecticut	Burr Creek	Documented	1980			Whitledge 1985
Connecticut	Connecticut River	Documented	1980		Episodic events in tidal freshwater zone.	Whitledge 1985, Bricker et al. 2007
Connecticut	Housatonic River	None	1980			Whitledge 1985
Connecticut	Johnson Creek	Concern	1980			Whitledge 1985
Connecticut	Naugatuck River	Concern	1980			Whitledge 1985
Connecticut	Pawcatuck River	Concern	1980			Whitledge 1985
Connecticut	Quinnipiac River	Concern	1980			Whitledge 1985
Connecticut	Thames River	None	1980			Whitledge 1985
Connecticut	Yellow Mill Channel	Concern	1980			Whitledge 1985
Delaware	Bald Eagle Creek	Documented	2000	Fish kills.	Eutrophication and limited tidal exchange with adjacent body of water.	Luther et al. 2004
Delaware	Blackwater Landing	Documented	1990			Sanger et al. 2002
Delaware	Dead-end canals/ Inland bays	Documented	1990		Dead-end canals have limited tidal exchange with adjacent body of water.	Maxted et al. 1997
Delaware	Delaware Bay	None	1980			Whitledge 1985

Table A1. (Continued)

State	System	Condition	Decade Condition First Recorded	Faunal Response	Trends and Other Comments	Reference
Delaware	Delaware River	Documented	1910	Recovering American shad and striped bass fishery.	Low dissolved oxygen first recorded in 1915. Recovered from low dissolved oxygen through nutrient management.	Weisberg et al. 1996, Summers et al. 1997, Patrick 1988
Delaware	Indian River Bay	Documented	1980			Whitledge 1985
Delaware	Little Assawoman Bay	Concern	1980	Fish kills.		Whitledge 1985, Bricker et al. 2007
Delaware	Murderkill River	Documented	1960		Caused by sewage discharge.	deWitt and Daiber 1974
Delaware	Pepper Creek	Documented	2000	Fish migrate in and out of creek depending on dissolved oxygen levels.	Daily cycle of dissolved oxygen from supersaturated to hypoxic.	Tyler 2004, Tyler and Targett 2007
Delaware	Rehoboth Bay	Documented	1980			Whitledge 1985
Maryland	Anacostia River	Concern	1980			Whitledge 1985
Maryland	Assawoman Bay	Documented	1980		No hypoxia in 1970s.	Whitledge 1985, Bricker et al. 2007, Boynton et al. 1996
Maryland	Back Creek	None	1980			Whitledge 1985
Maryland	Back River	Documented	1980			Whitledge 1985
Maryland	Baltimore Harbor	Documented	1980			Whitledge 1985
Maryland	Big Annemessex River	None	1980			Whitledge 1985, CBP 2008
Maryland	Bird River	None	1980			Whitledge 1985
Maryland	Bodkin Creek	Documented	1980			Whitledge 1985
Maryland	Bohemia	None	1990			CBP 2008
Maryland	Bush River	None	1980			Whitledge 1985

State	System	Condition	Decade Condition First Recorded	Faunal Response	Trends and Other Comments	Reference
Maryland	Cabin John Creek	Concern	1980			Whitledge 1985
Maryland	Chesapeake Bay Mainstem	Documented	1930	Low dissolved oxygen killed crabs in crab pots. Mortality of benthos with annual recolonization.	Seasonal anoxia detected in sediment record as far back as 1934-1948. Low dissolved oxygen events in the 1800s. Low dissolved oxygen occurrence has increased from 1980 to 2000, area of hypoxia and anoxia has increased.	Zimmerman and Canual 2000, Newcombe and Horne 1938, Officer et al. 1984, Seliger et al. 1985, Holland et al., 1987, Boesch et al. 2001, Hagy et al. 2004, Kemp et al. 2005
Maryland	Chester River	Documented	1980	Avoidance and some mortality of fishes. Mortality of benthos with annual recolonization.	Low dissolved oxygen events occur in June-Aug. Dissolved oxygen declined from 1990s to 2000s	Whitledge 1985, Bricker et al. 1999, 2007. CBP 2008
Maryland	Chincoteague Bay	Concern	1980		No hypoxia in 1970s.	Whitledge 1985, Bricker et al. 2007, Boynton et al. 1996
Maryland	Choptank River	Documented	1980		Low dissolved oxygen events occur in June-Sept.	Whitledge 1985, Bricker et al. 2007
Maryland	Corsica River	Documented	2000	Fish kill from combined stress of HAB toxins and low dissolved oxygen.		Bricker et al. 2007

Table A1. (Continued)

State	System	Condition	Decade Condition First Recorded	Faunal Response	Trends and Other Comments	Reference
Maryland	Dead-end canals	Documented	1990		Dead-end canals have limited tidal exchange with adjacent body of water.	Maxted et al. 1997
Maryland	Eastern Bay	None	1990			CBP 2008
Maryland	Elk River	None	1980			Whitledge 1985
Maryland	Fishing Bay	None	1990			CBP 2008
Maryland	Gunpowder River	None	1980			Whitledge 1985, CBP 2008
Maryland	Honga	None	1990			CBP 2008
Maryland	Isle of Wight Bay	Documented	1980		No hypoxia in 1970s.	Whitledge 1985, Bricker et al. 2007, Boynton et al. 1996
Maryland	Jug Bay	Documented	1990			Sanger et al. 2002
Maryland	Little Choptank	None	1990			CBP 2008
Maryland	Little River/ Browns Creek	None	1980			Whitledge 1985
Maryland	Magothy	Documented	1990			CBP 2008
Maryland	Manokin	None	1990			CBP 2008
Maryland	Manokin River	Concern	1980			Whitledge 1985
Maryland	Mattawoman Creek	None	1980			Whitledge 1985, CBP 2008
Maryland	Middle	None	1990			CBP 2008
Maryland	Miles River	None	1980			Whitledge 1985
Maryland	Nanjemoy Creek	None	1980			Whitledge 1985
Maryland	Nanticoke River	None	1980			Whitledge 1985
Maryland	Nassawango Creek	None	1980			Whitledge 1985
Maryland	Newport Bay	None	1980		No hypoxia in 1970s.	Whitledge 1985, Bricker et al. 2007, Boynton et al. 1996

State	System	Condition	Decade Condition First Recorded	Faunal Response	Trends and Other Comments	Reference
Maryland	Northeast River	None	1980			Whitledge 1985
Maryland	Oxon Run	Concern	1980			Whitledge 1985
Maryland	Patapsco River	Documented	1990			Riedel et al. 1999, CBP 2008
Maryland	Patuxent River	Documented	1930	Avoidance, low egg hatching/ larval mortality, bay anchovy eggs killed by hypoxia. Mortality of benthos with annual recolonization.	Duration and spatial extent of low dissolved oxygen has increased from 1930-79. Dissolved oxygen declined from 1990s to 2000s. Indications that hypoxia occurred in 1700s from watershed modifications and wet periods. Rise in anoxia in 1970s linked to nutrient runoff and wet weather.	Keister et al. 2000, Breitburg et al. 1997, 1997, Bricker et al. 1999, 2007, Cronin and Vann 2003
Maryland	Piscataway Creek	Concern	1980			Whitledge 1985
Maryland	Pocomoke River	None	1980			Whitledge 1985
Maryland	Port Tobacco River	Concern	1980			Whitledge 1985
Maryland	Potomac River	Documented	1910	Mortality of benthos with annual recolonization.	Low dissolved oxygen in May-Sept; increased spatial coverage of hypoxia from 1910s. Some improvement in dissolved oxygen levels from 1990s to 2000s.	Sale and Skinner 1917, Whitledge 1985, Bricker et al. 1999, 2007, CBP 2008
Maryland	Rhode River	None	1990			CBP 2008

Table A1. (Continued)

State	System	Condition	Decade Condition First Recorded	Faunal Response	Trends and Other Comments	Reference
Maryland	Rock Creek	Concern	1980			Whitledge 1985
Maryland	Sassafras River	None	1980			Whitledge 1985
Maryland	Seneca Creek	Concern	1980			Whitledge 1985
Maryland	Severn River	Documented	1990			CBP 2008
Maryland	Sinepuxent Bay	Concern	1980		No hypoxia in 1970s.	Whitledge 1985, Bricker et al. 2007, Boynton et al. 1996
Maryland	St. Leonard Creek	Documented	1990			Summers et al. 1997
Maryland	St. Martin River	None	1970		No hypoxia in 1970s.	Boynton et al. 1996
Maryland	St. Marys River	Concern	1980			Whitledge 1985
Maryland	Susquehanna River	Concern	1980			Whitledge 1985
Maryland	Swan Creek	None	1980			Whitledge 1985
Maryland	Tangier Sound	Concern	1990			Rabalais 1998, Bricker et al. 2007
Maryland	Turville Creek	None	1970		No hypoxia in 1970s.	Boynton et al. 1996
Maryland	Wicomico River	None	1990			CBP 2008
Maryland	Wye River	None	1980			Whitledge 1985
Massachusetts	Acushnet River	Concern	1970	Fish kill, maybe dissolved oxygen related August 1973.		Whitledge 1985
Massachusetts	Apponaganset Bay	None	1970			Whitledge 1985
Massachusetts	Buzzards Bay	Documented	1980		Summer hypoxia.	Whitledge 1985, Bricker et al. 2007, Buzzards Bay NEP 2008
Massachusetts	Clarks Cove	None	1970			Whitledge 1985
Massachusetts	New Bedford Harbor	Documented	1930	Loss of fishes and shellfish.	Low dissolved oxygen may have occurred in 1800s associated with whaling and textile	Whitledge 1985, NOAA 2008

State	System	Condition	Decade Condition First Recorded	Faunal Response	Trends and Other Comments	Reference
					industries, by 1930 water quality poor.	
Massachusetts	Tauton River	Documented	1970			Whitledge 1985
New Jersey	Barnegat Bay	Documented	1980	Mass mortality of benthos with annual recolonization.		Whitledge 1985, Moser 1998
New Jersey	Great Egg Harbor River	Documented	1980			Whitledge 1985, Glenn et al. 1996
New Jersey	Inland Bays	Concern	2000			Bricker et al. 2007
New Jersey	Little Egg Inlet	Documented	1970			Garlo et al. 1979
New Jersey	Metedeconk River	None	1980			Whitledge 1985
New Jersey	Mullica River Estuary	Documented	1990		Dissolved oxygen low in 1980s, hypoxic in 1990s.	Whitledge 1985, Glenn et al. 1996
New Jersey	Newark Bay	Concern	1980			Whitledge 1985
New Jersey	Passaic River	Concern	1980			Whitledge 1985
New Jersey	Raritan Bay	Documented	1970		Some improvement in dissolved oxygen from 1990s to 2000s. Improved from nutrient management.	Christensen and Packard 1976, Whitledge 1985, Bricker et al. 1999, 2007
New Jersey	Raritan River	Documented	1970			Whitledge 1985
New Jersey	Sandy Hook Bay	None	1980			Whitledge 1985
New Jersey	Toms River	None	1980			Whitledge 1985
New Jersey	Townsend- Hereford Inlet	Documented	1990			Glenn et al. 1996, 2004
New Jersey	Tuckahoe River	None	1980			Whitledge 1985
New Jersey	Wading River	Concern	1980			Whitledge 1985
New York	Arthur Kill	Documented	1970			Whitledge 1985
New York	Berger Basin	Documented	1970			Rhoads et al. 2001
New York	Carmans River	Documented	1990		Diel hypoxia present at times.	Zaikowski et al. 2008

Table A1. (Continued)

State	System	Condition	Decade Condition First Recorded	Faunal Response	Trends and Other Comments	Reference
New York	East River	Documented	1920		Decline in dissolved oxygen recorded in 1910-1930. Now dissolved oxygen improved from nutrient management.	Parker and O'Reilly 1991, Whitledge 1985
New York	Flushing Bay	Documented	1990			Rhoads et al. 2001
New York	Gardiners Bay	None	1980			Whitledge 1985
New York	Grassy Bay	Documented	1970		Areas anoxic in June-August.	Rhoads et al. 2001, Bricker et al. 2007
New York	Great South Bay	Documented	1980		Improved, but tidal creeks periodically hypoxic between July and September. Diel hypoxia may occur.	Whitledge 1985, Bricker et al. 2007, Zaikowski et al. 2008
New York	Hackensack	Concern	1980			Whitledge 1985
New York	Harlem River	Documented	1970			Whitledge 1985
New York	Hudson River	Documented	1960		Recovered from low dissolved oxygen through nutrient management.	Bronsnan and O'Shea 1996, Whitledge 1985
New York	Kill van Kull River	Documented	1970			Whitledge 1985
New York	Long Island Sound	Documented	1970	Avoidance and some mortality of fishes. About 30% of Sound benthos impacted by hypoxia.	Dissolved oxygen declined from 1990s to 2000s. Area of hypoxia has increased from 1980s. Hypoxic June to September; duration, frequency, spatial	Howell and Simpson 1994, Welsh et al. 1994, Schimmel et al. 1999, Anderson and Taylor 2001, Lee and Lwiza 2007

State	System	Condition	Decade Condition First Recorded	Faunal Response	Trends and Other Comments	Reference
					coverages all increased since 1987.	
New York	Lower Bay	Concern	1980			Whitledge 1985
New York	Mill Basin	Documented	1990	Avoidance by mobile fauna.		Rhoads et al. 2001
New York	Mohawk River	Concern	1980			Whitledge 1985
New York	Narrows	Documented	1970			Whitledge 1985, Lee and Lwiza 2007
New York	New York Bight	Documented	1970	Surf clam/finfish mortality, Avoidance, northward migration of bluefish blocked. Mass mortality of benthos with multi-year recovery. Large economic losses.	Calm weather led to stratification and bloom of Ceratium tripos, one time event. May have involved upwelling on nutrient-rich water.	Segar and Berberian 1976, Garlo et al. 1979, Sindermann and Swanson 1980, Swanson and Parker 1988, Figley et al. 1979, Azarovitz et al. 1979, Boesch and Rabalais 1991
New York	New York Bight Sewage Dump	Documented	1970		Summer hypoxia associated with sewage dump site, 10% saturation at center of dump site.	Segar and Berberian 1976
New York	New York City Harbor	Documented	1920	Stressed fishes. Mass mortality of benthos with annual recolonization.	Decline in dissolved oxygen recorded in 1910-1930. Now dissolved oxygen improved from nutrient management.	Parker and O'Reilly 1991
New York	Norton Basin	Documented	1990	Avoidance by mobile fauna.		Rhoads et al. 2001

Table A1. (Continued)

State	System	Condition	Decade Condition First Recorded	Faunal Response	Trends and Other Comments	Reference
New York	Patchogue River	Documented	1990	Benthic species richness severely impacted.	Channelization induced stratification an important factor. Deep areas hypoxic year round.	Zaikowski et al. 2008
New York	Shellbank Basin	Documented	1990			Rhoads et al. 2001
New York	Swan River	Documented	1990	Benthic species richness moderately to severely impacted.	Channelization induced stratification an important factor.	Zaikowski et al. 2008
New York	Tivolie South	Documented	1990			Sanger et al. 2002
New York	Upper Bay	Concern	1980			Whitledge 1985
Rhode Island	Greenwich Bay	Documented	2000	Fish kill in 2003.	0-4% of area <2.3 mg/l dissolved oxygen.	Deacutis et al. 2006, Melrose et al. 2007
Rhode Island	Mount Hope Bay	Documented	2000		0.5-2.5% of area <2.3 mg/l dissolved oxygen.	Deacutis et al. 2006, Melrose et al. 2007
Rhode Island	Narragansett Bay	Documented	1950	Reduced abundance or absence of benthos.	Hypoxic June to September, related to high nutrient loads. Hypoxia and anoxia present in 1950s in upper part of bay, low dissolved oxygen widespread by 1980s. Dissolved oxygen declined from 1990s to 2000s.	Whitledge 1985, Bricker et al. 1999, 2007, Melrose et al. 2007, Bergondo et al. 2005, Deacutis et al. 2006, Cicchetti et al. 2006
Rhode Island	Pettaquamscutt River	Documented	1990			Wilkin and Barnes 1997

State	System	Condition	Decade Condition First Recorded	Faunal Response	Trends and Other Comments	Reference
Rhode Island	Potters Cove	Documented	1990			Sanger et al. 2002
Rhode Island	Providence River	Documented	1980		Related to sewage discharges, 27-39% of the river was <2.3 mg/l dissolved oxygen.	Whitledge 1985, Bergondo et al. 2005, Deacutis et al. 2006, Melrose et al. 2007
Virginia	Appomattox River	None	1990			CBP 2008
Virginia	Back Bay	None	1980			Whitledge 1985
Virginia	Chesapeake Bay, lower	Documented	1980	Lower benthic diversity, lower biomass, lower proportion of deep-dwelling biomass at hypoxia-affected areas.		Dauer et al 1992
Virginia	Chickahominy	None	1990			CBP 2008
Virginia	Chuckatuck River	None	1990			CBP 2008
Virginia	Cobb Bay	None	1980			Whitledge 1985
Virginia	Corrotoman River	None	1990			CBP 2008
Virginia	East River	None	1990			CBP 2008
Virginia	Elizabeth River, Eastern Branch	Documented	1990			CBP 2008
Virginia	Elizabeth River, Southern Branch	Documented	1990			CBP 2008
Virginia	Elizabeth River, Western Branch	Concern	1990			CBP 2008
Virginia	Goodwin Island	Documented	1990		Low dissolved oxygen in submerged aquatic vegetation bed.	Sanger et al. 2002
Virginia	Grays Creek	None	1990			CBP 2008
Virginia	Hog Island Bay	None	1980			Whitledge 1985

Table A1. (Continued)

State	System	Condition	Decade Condition First Recorded	Faunal Response	Trends and Other Comments	Reference
Virginia	James River	None	1980		Dissolved oxygen improved from nutrient management.	Whitledge 1985
Virginia	Jones-Gilligan Creek	None	1990			CBP 2008
Virginia	Lafayette River	Concern	1990			CBP 2008
Virginia	Lynnhaven Inlet	None	1990			CBP 2008
Virginia	Mattaponi River	Documented	1990			CBP 2008
Virginia	Mobjack Bay	Documented	1990			CBP 2008
Virginia	Mud Creek	None	1990			CBP 2008
Virginia	Nansemond River	Documented	1990			CBP 2008
Virginia	North River	None	1990			CBP 2008
Virginia	Onancock Creek	Documented	2000			Wang 2005
Virginia	Pagan River	None	1990			CBP 2008
Virginia	Pamunkey River	None	1990			CBP 2008
Virginia	Paradise Creek	None	1990			CBP 2008
Virginia	Piankatank River	None	1990			CBP 2008
Virginia	Pocomoke Sound	None	1980			Whitledge 1985
Virginia	Poropotank River	None	1990			CBP 2008
Virginia	Queen Creek	None	1990			CBP 2008
Virginia	Rappahannock River	Documented	1980	Avoidance by mobile fauna. Mortality of benthos with annual recolonization.	Some anoxia in summer.	Llansó 1992, Kuo et al. 1991
Virginia	Robinson Creek	None	1990			CBP 2008
Virginia	Sarah Creek	None	1990			CBP 2008
Virginia	Scuffletown Creek	None	1990			CBP 2008
Virginia	Severn Creek	None	1990			CBP 2008
Virginia	South Bay	None	1980			Whitledge 1985
Virginia	Taskinas Creek	Documented	1990			Sanger et al. 2002

State	System	Condition	Decade Condition First Recorded	Faunal Response	Trends and Other Comments	Reference
Virginia	Timberneck Creek	None	1990			CBP 2008
Virginia	Totuskey Creek	None	1990			CBP 2008
Virginia	Urbanna Creek	None	1990			CBP 2008
Virginia	Ware River	None	1990			CBP 2008
Virginia	Warwick River	None	1990			CBP 2008
Virginia	Weeks Creek	None	1990			CBP 2008
Virginia	Willoughby Bay	None	1990			CBP 2008
Virginia	York River	Documented	1970	Avoidance by mobile fauna with reduced diversity following low dissolved oxygen events. Little to no response in the benthos.	Hypoxia associated with spring-neap tidal cycle.	Haas 1977, Pihl et al. 1991, 1992, Diaz et al. 1992, Sagasti et al 2001, 2003, Seitz et al. 2003
		Northeast Region				
Maine	Androscoggin River	None	1980			Whitledge 1985, Bricker et al. 2007
Maine	Blue Hill Bay	None	1980			Whitledge 1985
Maine	Casco Bay	None	1980			Whitledge 1985, Bricker et al. 2007
Maine	Cobscook Bay	None	1980			Whitledge 1985
Maine	Damariscotta River	None	1980			Whitledge 1985
Maine	Englishman Bay	None	1980			Whitledge 1985
Maine	Fore River	None	1980			Whitledge 1985
Maine	Frenchman Bay	None	1980			Whitledge 1985
Maine	Great Bay	Concern	2000		Evidence of decline in dissolved oxygen levels.	Bricker et al. 2006
Maine	Kennebec River	None	1980			Whitledge 1985
Maine	Machias Bay	None	1980			Whitledge 1985
Maine	Machias River Estuary	None	1980			Whitledge 1985

Table A1. (Continued)

State	System	Condition	Decade Condition First Recorded	Faunal Response	Trends and Other Comments	Reference
Maine	Muscongus Bay	None	1980			Whitledge 1985
Maine	Narraguagus Bay	None	1980			Whitledge 1985
Maine	Narraguagus River	Concern	1980			Whitledge 1985
Maine	Passamaquoddy Bay	None	1980			Whitledge 1985
Maine	Penobscot Bay/ River	None	1980			Whitledge 1985
Maine	Pleasant Bay	None	1980			Whitledge 1985
Maine	Saco Bay	None	1980			Whitledge 1985
Maine	Sheepscot Bay	None	1980			Whitledge 1985
Maine	St. Croix River	None	1980			Whitledge 1985
Maine	Wells Bay	Documented	1990			Sanger et al. 2002
Maine	Wells Inlet	Documented	1990		Daily cycle of dissolved oxygen from supersaturated to hypoxic, Submerged aquatic vegetation bed.	Sanger et al. 2002
Maine	Wohoa Bay	None	1980			Whitledge 1985
Massachusetts	Boston Harbor	None	1980		Entire region recovered from low dissolved oxygen through nutrient management.	Whitledge 1985, Bricker et al. 1999, 2007, Diaz et al. 2008
Massachusetts	Cape Cod Bay	Documented	1990		Blooms.	Whitledge 1985, Rabalais 1998
Massachusetts	Charles River	Documented	1980			Taylor 2000
Massachusetts	Dorchester Bay	Documented	1980		Had poor water quality in 1980s. Low dissolved oxygen in 1980s now improved from nutrient management.	Maciolek et al. 2005, Diaz et al. 2008

State	System	Condition	Decade Condition First Recorded	Faunal Response	Trends and Other Comments	Reference
Massachusetts	Duxbury Bay	None	1980			Whitledge 1985
Massachusetts	Fore River	Documented	1980			Neponset River Watershed Association 2004
Massachusetts	Herring River	Documented	1980	Fish kills, decline of alewife fishery.	Anoxia, 1-3 weeks.	Portnoy 1991
Massachusetts	Massachusetts Bay	None	1990			Bricker et al. 2007
Massachusetts	Merrimack River	Documented	1970			Whitledge 1985
Massachusetts	Plum Island Sound	None	1990			Rabalais 1998, Bricker et al. 2007
Massachusetts	Waquoit Bay	Documented	1990		Daily cycle of dissolved oxygen from saturated to hypoxic.	Fritz et al. 1996, D'Avanzo and Kremer 1994, Bricker et al. 2007
New Hampshire	Great Bay	Documented	1980			Whitledge 1985, Bricker et al. 2007
New Hampshire	Hampton Harbor Estuary	None	1990			Rabalais 1998
New Hampshire	Piscataqua River	None	1980			Whitledge 1985, Bricker et al. 2007
		South Atlantic Region				
Florida	Amelia River	Documented	1970			Windsor 1985
Florida	Banana River	Concern	1980			Windsor 1985
Florida	Biscayne Bay	Documented	1970		Related to releases of freshwater from canals. Dissolved oxygen declined from 1990s to 2000s.	Bricker et al. 1999, 2007, Windsor 1985, Leverone 1995
Florida	Canal 17	Documented	1980			Whitledge 1985
Florida	Cedar Creek	Documented	1980			Windsor 1985

Table A1. (Continued)

State	System	Condition	Decade Condition First Recorded	Faunal Response	Trends and Other Comments	Reference
Florida	East Holloway Canal	Documented	1980			Windsor 1985
Florida	Eau Gallie Harbor	Concern	1980			Whitledge 1985
Florida	Gulf and West Canals	Documented	1980			Windsor 1985
Florida	Halifax River	Documented	1980			Windsor 1986
Florida	Haw Creek	Documented	1980			Windsor 1985
Florida	Hillsboro Canal	Documented	1980			Windsor 1985
Florida	Indian River	Documented	2000		Improvement in dissolved oxygen levels from 1990s to 2000s.	Bricker et al. 1999, 2007, Windsor 1985
Florida	Lofton Creek	Concern	1980			Windsor 1985
Florida	Loxahatchee River	Concern	1980			Windsor 1985
Florida	Matanzas River	Concern	1980			Windsor 1985
Florida	Miami Canal	Documented	1980			Windsor 1985
Florida	Mosquito Lagoon	Concern	1980			Windsor 1985
Florida	Naples Bay	Documented	1980			Windsor 1985
Florida	Nassau River	None	1980			Windsor 1985
Florida	North and South New River Canals	Documented	1980			Windsor 1985
Florida	North River	Documented	1980			Windsor 1987
Florida	Ortega River	Documented	1980			Windsor 1985
Florida	Pottsburg Creek	Documented	1980			Windsor 1985
Florida	Ribault River	Documented	1980			Windsor 1985
Florida	Rice Creek	Documented	1980			Windsor 1985
Florida	Sebastian Creek	Documented	1980			Windsor 1985
Florida	Sisters Creek	Concern	1980			Windsor 1985
Florida	South Amelia River	Concern	1980			Windsor 1985
Florida	South Relief Canal	Documented	1980			Windsor 1985
Florida	Spruce Creek	None	1980			Windsor 1985
Florida	St. Johns River	Documented	1970	Mortality of benthos,		Windsor 1985,

State	System	Condition	Decade Condition First Recorded	Faunal Response	Trends and Other Comments	Reference
				massive fish kills with unknown cause, may be multiple stressors.		Mason 1998
Florida	St. Lucie Canal	Concern	1980			Windsor 1985
Florida	St. Lucie River	Documented	1980			Chamberlain and Hayward 1996, Windsor 1985
Florida	St. Marys River	Documented	1970	Anoxia develops in summer.	No significant declining trend in dissolved oxygen from 1960s to 2000s	Windsor 1985, Verity et al. 2006
Florida	Sykes Creek	Documented	1980			Windsor 1985
Florida	Tenmile Creek	Documented	1980			Windsor 1985
Florida	Tolomato River	Concern	1980			Windsor 1985
Florida	Tomoka River	Concern	1980			Windsor 1985
Florida	Trout River	Documented	1980			Windsor 1985
Florida	Turkey Creek	Documented	1980			Windsor 1986
Florida	West Palm Beach Canal	Documented	1980			Windsor 1985
Florida	Whitewater Bay	Concern	1980			Windsor 1985
Georgia	Altamaha Sound	Concern	2000		Declining trend in dissolved oxygen related to nutrient loadings. No hypoxia in 1980s.	Stanley 1985, Verity et al. 2006
Georgia	Altamaha River	Documented	2000		Significant declining trend in dissolved oxygen from 1970s to 2000s. No hypoxia in 1980s.	Stanley 1985, Verity et al. 2006
Georgia	Brunswich River	Concern	1970			Stanley 1985

Table A1. (Continued)

State	System	Condition	Decade Condition First Recorded	Faunal Response	Trends and Other Comments	Reference
Georgia	Brunswick Harbor	Concern	1970			Stanley 1985
Georgia	Cumberland Sound	Documented	2000		Declining trend in Dissolved oxygen related to nutrient loadings.	Stanley 1985, Verity et al. 2006
Georgia	Doboy Sound	None	1980			Stanley 1985
Georgia	Duplin River	Documented	1970		Daily cycle of dissolved oxygen from supersaturated to hypoxic.	Frankenberg 1976, Stanley 1985
Georgia	Medway River	None	1970			Stanley 1985
Georgia	Ogeechee River	Documented	2000		Significant declining trend in dissolved oxygen from 1960s to 2000s. No hypoxia in 1970s.	Stanley 1985, Verity et al. 2006
Georgia	Ogeechee Sound	Concern	2000		Declining trend in dissolved oxygen related to nutrient loadings.	Stanley 1985, Verity et al. 2006
Georgia	Ossabaw Sound	Documented	1980		Related to organic matter washing out of swamps and marshes. Dissolved oxygen declined from 1990s to 2000s.	Stanley 1985, Bricker et al. 1999, 2007
Georgia	Sapelo Island	Documented	2000		Daily cycle of dissolved oxygen from supersaturated to	Sanger et al. 2002

State	System	Condition	Decade Condition First Recorded	Faunal Response	Trends and Other Comments	Reference
					hypoxic.	
Georgia	Sapelo Sound	Concern	1990		No hypoxia in 1980s, declining trend is dissolved oxygen for 2000s.	Stanley 1985, Verity et al. 2006
Georgia	Satilla River	Documented	2000		Significant declining trend in dissolved oxygen from 1960s to 1990s. Not hypoxic in 1970s.	Stanley 1985, Verity et al. 2006
Georgia	Savannah River	Documented	1980		Nonpoint sources a factor. Significant declining trend in dissolved oxygen from 1960s to 2000s.	Stanley 1985, Verity et al. 2006, Bricker et al. 2007
Georgia	Skidaway River Estuary	Concern	1990		High rate of microbial respiration causes low dissolved oxygen despite strong vertical and horizontal mixing. Declining trend in dissolved oxygen related to nutrient loadings.	Verity et al. 2006, Bricker et al. 2007
Georgia	St. Andrews Sound	Concern	1990		No hypoxia in 1980s, declining trend is dissolved oxygen for 2000s.	Stanley 1985, Verity et al. 2006
Georgia	St. Catherine Sound	None	1980			Stanley 1985
Georgia	St. Marys River	Concern	1970			Stanley 1985
Georgia	St. Simons Sound	None	1980			Stanley 1985, Rabalais 1998, Bricker et al. 2007

Table A1. (Continued)

State	System	Condition	Decade Condition First Recorded	Faunal Response	Trends and Other Comments	Reference
Georgia	Turtle River	Documented	1970			Stanley 1985
Georgia	Tybee Creek	None	1970			Stanley 1985
Georgia	Wassaw Sound	None	1980			Stanley 1985
Georgia	Wilmington River	None	1970			Stanley 1985
North Carolina	Ace Basin	Documented	1990			Sanger et al. 2002
North Carolina	Albemarle Sound	Documented	1970	Mortality of benthos and mobile fauna.	Hurricane increased freshwater and nutrient input. Hypoxia has worsened in 2000s.	Stanley 1985, Bricker et al. 1999, 2007, Paerl et al. 2000
North Carolina	Alligator River	None	1980			Stanley 1985
North Carolina	Barnards Creek	Documented	2000		Combination of agricultural, urban, suburban runoff contribute to nutrient loading.	Mallin et al. 2006
North Carolina	Black River	Documented	1990		Combination of agricultural, urban, suburban runoff contribute to nutrient loading.	Mallin et al. 2006
North Carolina	Bogue Sound	Concern	1980			Stanley 1985
North Carolina	Browns Creek	Concern	2000		Combination of agricultural, urban, suburban runoff contribute to nutrient loading.	Mallin et al. 2006
North Carolina	Cape Fear River	Documented	1970	Fish kills, reduced benthos.		Stanley 1985, Mallin et al. 1999, 2006, Posey

State	System	Condition	Decade Condition First Recorded	Faunal Response	Trends and Other Comments	Reference
						et al. 1999
North Carolina	Carolina Beach Lake	Documented	2000		Combination of agricultural, urban, suburban runoff contribute to nutrient loading.	Mallin et al. 2006
North Carolina	Chowan River	Documented	1980			Stanley 1985
North Carolina	Colly Creek	Concern	2000		Combination of agricultural, urban, suburban runoff contribute to nutrient loading.	Mallin et al. 2006
North Carolina	Core Sound	None	1980			Stanley 1985
North Carolina	Currituck Sound	None	1980			Stanley 1985
North Carolina	Dutchman Creek	Concern	1970			Stanley 1985
North Carolina	Futch Creek	Documented	2000		Combination of agricultural, urban, suburban runoff contribute to nutrient loading.	MacPherson et al. 2007, Mallin et al. 2006
North Carolina	Great Coharie Creek	Documented	2000		Combination of agricultural, urban, suburban runoff contribute to nutrient loading.	Mallin et al. 2006
North Carolina	Hammond Creek	Documented	2000		Combination of agricultural, urban, suburban runoff contribute to nutrient loading.	Mallin et al. 2006

Table A1. (Continued)

State	System	Condition	Decade Condition First Recorded	Faunal Response	Trends and Other Comments	Reference
North Carolina	Hewletts Creek	Documented	2000		Heavily developed watershed.	MacPherson et al. 2007, Mallin et al. 2006
North Carolina	Little Coharie Creek	Concern	2000		Combination of agricultural, urban, suburban runoff contribute to nutrient loading.	Mallin et al. 2006
North Carolina	Little River	None	1980			Stanley 1985
North Carolina	Masonboro Inlet	Documented	1990			Sanger et al. 2002
North Carolina	Motts Creek	Documented	2000		Combination of agricultural, urban, suburban runoff contribute to nutrient loading.	Mallin et al. 2006
North Carolina	Neuse River Estuary	Documented	1980	Fish kills, mortality of oyster. Reduced prey abundance and habitat quality for fish. Mortality of benthos with decreased abundance.	Hypoxia may have developed earlier based on Macoma populations. Improvement in dissolved oxygen from 1990s to 2000s.	Paerl et al. 1995, 1998, Lenihan and Peterson 1998, Lenihan 1999, Whitledge 1985, Buzzelli et al. 2002, Bishiop et al. 2006, Bell and Eggleston 2005, Baird et al. 2004, Aumann et al. 2006, Powers et al. 2005
North Carolina	New River	Documented	1980			Stanley 1985, Bricker et al. 2007
North Carolina	Newport River	None	1980			Stanley 1985
North Carolina	North River	None	1980			Stanley 1985
North Carolina	Northeast Cape Fear River	Documented	1990		Combination of	Mallin et al. 2006

State	System	Condition	Decade Condition First Recorded	Faunal Response	Trends and Other Comments	Reference
					agricultural, urban, suburban runoff contribute to nutrient loading.	
North Carolina	Pages Creek	Documented	2000		Combination of agricultural, urban, suburban runoff contribute to nutrient loading.	MacPherson et al. 2007, Mallin et al. 2006
North Carolina	Pamlico River	Documented	1960	Mortality of fishes, mass mortality of benthos, low macrobenthic diversity and density in summer, recolonization by winter.		Stanley 1985, Tenore 1972, Hobbie et al. 1975, Stanley and Nixon 1992
North Carolina	Pamlico Sound	Documented	1990	Common finfish species (pinfish, spot, croaker) had skin lesions and signs of systemic bacterial infections.	No hypoxia in 1960s through the 1980s, low dissolved oxygen developed in late 1990s. Hurricane increased freshwater and nutrient input.	Paerl et al. 2000, Stanley 1985
North Carolina	Pasquotank River	None	1980			Stanley 1985
North Carolina	Perquimans River	None	1980			Stanley 1985
North Carolina	Pungo River	None	1980			Stanley 1985
North Carolina	Six Runs Creek	Documented	2000		Combination of agricultural, urban, suburban runoff contribute to nutrient loading.	Mallin et al. 2006

Table A1. (Continued)

State	System	Condition	Decade Condition First Recorded	Faunal Response	Trends and Other Comments	Reference
North Carolina	Smith Creek	Documented	2000		Combination of agricultural, urban, suburban runoff contribute to nutrient loading.	Mallin et al. 2006
North Carolina	White Oak River	None	1980			Stanley 1985
South Carolina	Ashley River	None	1980			Stanley 1985
South Carolina	Beresford Creek	Documented	1990			Lerberg et al. 2000
South Carolina	Broad River	Documented	2000		Dissolved oxygen improved from 1990s to 2000s.	Bricker et al. 1999, 2007
South Carolina	Bull Creek	Documented	1990			Lerberg et al. 2000
South Carolina	Bulls Bay	None	1980			Stanley 1985
South Carolina	Charleston Harbor	Documented	2000		Improved wastewater treatment, phosphate ban. Improvement in dissolved oxygen from 1990s to 2000s.	Bricker et al. 1999, 2007, Windsor 1985
South Carolina	Combahee River	None	1980			Stanley 1985
South Carolina	Cooper River	Concern	1980			Whitledge 1985
South Carolina	Coosaw River	None	1980			Stanley 1985
South Carolina	Copper River	Concern	1980			Stanley 1985
South Carolina	Deep Creek	Documented	1990			Lerberg et al. 2000
South Carolina	Diesel Creek	Documented	1990			Lerberg et al. 2000
South Carolina	Foster Creek	Documented	1990			Lerberg et al. 2000
South Carolina	James Island Creek	Documented	1990	Avoidance by fishes, mortality of benthos.		Burnett 1997
South Carolina	Kiawah River	None	1980			Stanley 1985
South Carolina	Little River Inlet	None	1980			Stanley 1985

State	System	Condition	Decade Condition First Recorded	Faunal Response	Trends and Other Comments	Reference
South Carolina	Long Bay	Documented	2000	Mortality of flounder.	Recently documented hypoxia may be related to multiple factors including high chlorophyll a levels.	Koepfler et al. 2007
South Carolina	Malind Creek	Documented	2000			Gillett et al. 2005
South Carolina	Murrells Inlet	None	1980			Stanley 1985
South Carolina	New Market Creek	Documented	1990			Lerberg et al. 2000
South Carolina	North Edisto River	Concern	1980			Stanley 1985, Bricker et al. 2007
South Carolina	North Inlet	Documented	2000			Stanley 1985, Sanger et al. 2002
South Carolina	North Santee River	None	1980			Stanley 1985
South Carolina	Okatee Creek	Documented	2000			Gillett et al. 2005
South Carolina	Orange Grove Creek	Documented	1990			Lerberg et al. 2000
South Carolina	Pee Dee River	Concern	1980			Stanley 1985
South Carolina	Port Royal Sound	None	1980			Stanley 1985
South Carolina	Rathall Creek	Documented	1990			Lerberg et al. 2000
South Carolina	Santee River	Concern	1990			Rabalais 1998, Bricker et al. 2007
South Carolina	Shem Creek	Documented	1990			Lerberg et al. 2000
South Carolina	South Atlantic Bight	Concern	1970		Dissolved oxygen around 3 mg/l at 100 m depth in 1973- 74.	Stanley 1985
South Carolina	South Edisto River	None	1980			Stanley 1985
South Carolina	South Santee River	None	1980			Stanley 1985
South Carolina	St. Helena Sound	Documented	1980			Stanley 1985, Bricker et al. 2007
South Carolina	Stono River	Documented	1990		No hypoxia in 1980s, low dissolved oxygen developed in 1990s.	Stanley 1985, Bricker et al. 1999, 2007

Table A1. (Continued)

State	System	Condition	Decade Condition First Recorded	Faunal Response	Trends and Other Comments	Reference
South Carolina	Wando River	Concern	1980			Stanley 1985, Whitledge 1985
South Carolina	Winyah Bay	Documented	1980			Stanley 1985, Bricker et al. 2007
		Pacific Coast Region and Hawaii				
Alaska	Ward Cove	Documented	1950	Periodic fish kill in 1950s.	Related to pulp mill discharge 1954-1997, circulation is pore in cove, dissolved oxygen still low in areas.	Karna 2003
California	Alamitos Bay	Documented	2000			Rabalais 1998
California	Anaheim Bay to Balsa Chica	Concern	1990			Rabalais 1998
California	Azevedo Pond	Documented	1990		Diel cycling, low dissolved oxygen related to high primary production.	Beck and Bruland 2000
California	Big Lagoon	Documented	1980			Collias 1985, Engle et al. 1999
California	Coyote Creek	Documented	1970	Fisherman reported absence of fish and pelagic invertebrates, fish returned after hypoxia ended.	Sewage spill.	Cloern and Oremland 1983
California	Drakes Estero	None	1980			Collias 1985
California	Eel River	None	1990			Rabalais 1998, Bricker et al. 2007

State	System	Condition	Decade Condition First Recorded	Faunal Response	Trends and Other Comments	Reference
California	Elkhorn Slough	Documented	1990		Dissolved oxygen improved from 1990s to 2000s.	Bricker et al. 1999, 2007, Sanger et al. 2002
California	Elkhorn Slough, Upper Pond	Documented	2000			Bricker et al. 1999, 2007, Sanger et al. 2002
California	Humboldt Bay/ Arcata Bay	None	1980			Collias 1985
California	Klamath River	None	1990			Rabalais 1998, Bricker et al. 2007
California	Long Beach Harbor	Documented	1960			Whitledge 1985
California	Los Angeles Harbor	Documented	1950	Mass mortality of benthos with multi-year recovery.	Improved from nutrient management.	Reish 1955, 2000
California	Mission Bay	None	1990			Rabalais 1998, Bricker et al. 2007
California	Monteray Bay	Documented	2000		Combination of anthropogenic and natural factors.	Okey 2003
California	Morro Bay	None	1990			Rabalais 1998, Bricker et al. 2007
California	Newport Bay	Documented	1980		Dissolved oxygen improved from 1990s to 2000s.	Collias 1985, Rabalais 1998
California	North San Francisco Bay Estuary	Documented	2000	Fish Kill	Anoxia develops.	Lehman et al. 2004
California	Palos Verde Shelf	None	1970		Organic content of sediment enriched from sewage, but no hypoxia.	Smith et al. 2001, Weisberg (Personal Communication)
California	San Diego Bay	Documented	1980			Collias 1985, Rabalais 1998
California	San Francisco Bay North	Concern	1980			Collias 1985

Table A1. (Continued)

State	System	Condition	Decade Condition First Recorded	Faunal Response	Trends and Other Comments	Reference
California	San Francisco Bay South	Documented	1960		Hypoxia eliminated with sewage treatment plants. Recovered from low dissolved oxygen through nutrient management.	Nichols et al. 1986, Bricker et al. 2007
California	San Joaquin River	Documented	1960	Salmon migration blocked when dissolved oxygen <6 mg/l. High bacterial respiration and nitrification make dissolved oxygen low, stratification is weak.	Sewage discharge and nonpoint runoff are sources of organic matter, primary production also contributes. First <2 mg/l dissolved oxygen in 1972.	Lehman et al. 2004, Bricker et al. 2007, Jassby and Van Nieuwenhuyse 2005
California	San Pedro Bay	None	1990			Rabalais 1998, Bricker et al. 2007
California	Santa Monica Bay	Documented	1990			Rabalais 1998, Bricker et al. 2007
California	Smith River	None	1980			Collias 1985
California	Tijuana Estuary	Documented	2000		Dissolved oxygen improved from 1990s to 2000s.	Bricker et al. 2007, Sanger et al. 2002
California	Tomales Bay/ Bodega Harbor	Concern	1980			Collias 1985
Hawaii	O'ahu, Off Southern Shore	Documented	2000	Mortality of benthos with low species richness and abundance.	Fish aquaculture.	Lee et al. 2006
Oregon	Alsea River	None	1980			Collias 1985

State	System	Condition	Decade Condition First Recorded	Faunal Response	Trends and Other Comments	Reference
Oregon	California Current System	Documented	2000	Mortality of fishes and mass mortality of benthos.	Shifting wind patterns lead to extensive shallow shelf hypoxia. Related to upwelling and phytoplankton bloom. Hypoxia developed on inner shelf in 2000s.	Grantham et al. 2004, Chan et al. 2008
Oregon	Columbia River	None	1980			Collias 1985, Keefer et al. 2008
Oregon	Coos Bay	None	1980			Collias 1985
Oregon	Coquille River	None	1980			Collias 1985
Oregon	Nehalem River	None	1980			Collias 1985
Oregon	Nestucca Bay	None	1980			Collias 1985
Oregon	Netarts Bay	None	1980			Collias 1985
Oregon	Rogue River	Documented	1980			Collias 1985, Sanger et al. 2002
Oregon	Siletz Bay	None	1980			Collias 1985
Oregon	Siuslaw River	None	1980			Collias 1985
Oregon	South Slough	Documented	1990		Daily cycle of dissolved oxygen from supersaturated to hypoxic.	Sanger et al. 2002
Oregon	Tillamook Bay	None	1980			Collias 1985
Oregon	Umpqua River	Concern	1970		1970s intrusion of low dissolved oxygen (<2 mg/l). No hypoxia in 1980s.	Collias 1985, Brown et al. 2007

Table A1. (Continued)

State	System	Condition	Decade Condition First Recorded	Faunal Response	Trends and Other Comments	Reference
Oregon	Yaquina Bay	Documented	2000		No hypoxia in 1980s. Low dissolved oxygen water at times advected into estuary from shelf.	Collias 1985, Brown et al. 2007
Washington	Admiralty Inlet	Concern	1990			Washington State Dept. Ecol. 2002
Washington	Bellingham Bay	Concern	1990		No hypoxia in 1980s.	Collias 1985, Washington State Dept. Ecol. 2002
Washington	Budd Inlet	Documented	1980			Whitledge 1985, Washington State Dept. Ecol. 2002
Washington	Commencement Bay	Concern	1990			Washington State Dept. Ecol. 2002
Washington	Dana Passage	None	1990			Washington State Dept. Ecol. 2002
Washington	Discovery Bay	Documented	1990			Washington State Dept. Ecol. 2002
Washington	DRA002	None	1990			Washington State Dept. Ecol. 2002
Washington	ELD002	None	1990			Washington State Dept. Ecol. 2002
Washington	Elliott Bay	Documented	1990			Washington State Dept. Ecol. 2002
Washington	GOR001	None	1990			Washington State Dept. Ecol. 2002
Washington	Grays Harbor	Concern	1980			Collias 1985
Washington	Hood Canal	Documented	1930	Fish did not enter	Dissolved oxygen	Paulson et al. 1993,

State	System	Condition	Decade Condition First Recorded	Faunal Response	Trends and Other Comments	Reference
				dissolved oxygen minimum later in water column while invertebrates did.	declined from 1990s to 2000s.	Bricker et al. 2007, Parker-Stetter and Horne 2008
Washington	Lynch Cove	Documented	1980		Low dissolved oxygen all year round.	Whitledge 1985, Washington State Dept. Ecol. 2002
Washington	Oakland Bay	None	1990			Washington State Dept. Ecol. 2002
Washington	Padilla Bay	Documented	1990		No hypoxia in 1980s, Low dissolved oxygen in submerged aquatic vegetation bed.	Collias 1985, Sanger et al. 2002
Washington	Penn Cove	Documented	2000			Washington State Dept. Ecol. 2002
Washington	POD007	None	1990			Washington State Dept. Ecol. 2002
Washington	Port Gamble	Concern	1990			Washington State Dept. Ecol. 2002
Washington	Port Gardner	None	1980			Collias 1985
Washington	Port Orchard System	Concern	1980			Collias 1985
Washington	Port Susan	Documented	1980			Collias 1985
Washington	Possession Sound	Concern	1990			Washington State Dept. Ecol. 2002
Washington	Puget Sound	None	1990			Rabalais 1998, Bricker et al. 2007
Washington	Samish Bay	Concern	1980		No hypoxia in 1980s.	Collias 1985
Washington	Saratoga Passage	Documented	1990			Washington State Dept. Ecol. 2002
Washington	Sequim/Discovery Bays	Documented	1980			Collias 1985

Table A1. (Continued)

State	System	Condition	Decade Condition First Recorded	Faunal Response	Trends and Other Comments	Reference
Washington	Sinclair Inlet	Concern	1990			Washington State Dept. Ecol. 2002
Washington	Skagit Bay	Documented	1990			Rabalais 1998
Washington	South Puget Sound	Documented	1980			Collias 1985
Washington	Straight of Georgia	Concern	1990			Washington State Dept. Ecol. 2002
Washington	TOT001	None	1990			Washington State Dept. Ecol. 2002
Washington	West Point	Documented	1990			Washington State Dept. Ecol. 2002
Washington	Whidbey Basin	Documented	1980			Collias 1985, Rabalais 1998
Washington	Willapa Bay	None	1980			Collias 1985
Washington	Willapa River	Documented	1990			Washington State Dept. Ecol. 2002

REFERENCES

Abadie SW, Poirrier MA. 2000. Increased density of large Rangia clams in Lake Pontchartrain after the cessation of shell dredging. *J. Shellfish Res.* 19:481-485.

Ahrnsbrak WF, Wing MR. 1998. Wind-Induced Hypolimnion Exchange in Lake Ontario's Kingston Basin: Potential Effects on Oxygen. *J. Great Lakes Res.* 24:145-151.

Anderson TH, Taylor GT. 2001. Nutrient pulses, plankton blooms, and seasonal hypoxia in western Long Island Sound. *Estuaries* 24:228-243.

Applebaum S, Montagna P, Ritter C. 2005. Status and trends of dissolved oxygen in Corpus Christi Bay, Texas, USA. Environ. Monitor. *Assess.* 107:297-311.

Aumann CA, Eby LA, Fagan WF. 2006. How transient patches affect population dynamics: The case of hypoxia and blue crabs. *Ecol. Monogr.* 76:415-438.

Azarovitz TR, Byrne CJ, Silverman MJ, Freeman BL, Smith WG, Turner SC, Halgren BA, Festa PJ. 1979. Effects on finfish and lobster. In: Oxygen depletion and associated benthic moralities in New York Bight, 1976. R.L. Swanson and C.J. Sindermann (eds.). *NOAA Professional Paper* 11, pp. 295-314.

Baird D, Christian RR, Peterson CH, Johnson GA. 2004. Consequences of hypoxia on estuarine ecosystem function: Energy diversion from consumers to microbes. *Ecol. Appl.* 14:805-822.

Beck NG, Bruland KW. 2000. Diel biogeochemical cycling in a hyperventilating shallow estuarine environment. *Estuaries* 23:177-187.

Bell GW, Eggleston DB. 2005. Species-specific avoidance responses by blue crabs and fish to chronic and episodic hypoxia. *Mar. Biol.* 146:761-770.

Bergondo DL, Kester DR, Stoffel HE, Woods WL. 2005. Time-series observations during the low sub-surface oxygen events in Narragansett Bay during summer 2001. *Mar. Chem.* 97:90-103.

Bishop MJ, Powers SP, Porter HJ, Peterson CH. 2006. Benthic biological effects of seasonal hypoxia in a eutrophic estuary predate rapid coastal development. Estuarine, Coastal and Shelf Science 70:415-422.

Blumberg AF, Di Toro DM. 1990. Effects of climate warming on dissolved oxygen concentrations in Lake Erie. *Trans. Am. Fish. Soc.* 11:210-223.

Boesch DF, Rabalais NN. 1991. Effects of hypoxia on continental shelf benthos: comparisons between the New York Bight and the Northern Gulf of Mexico. In *"Modern and ancient continental shelf anoxia*" (R. V. Tyson and T. H.

Boesch DF, Brinsfield RB, Magnien RE. 2001. Chesapeake Bay eutrophication: scientific understanding, ecosystem restoration, and challenges for agriculture. *J. Environ. Qual.* 30:303-320.

Boynton WR, Hagy JD, Murray L, Stokes C, Kemp WM. 1996. A comparative analysis of eutrophication patterns in a temperate coastal lagoon. *Estuaries* 19:408-421.

Breitburg DL. 1992. Episodic hypoxia in Chesapeake Bay: interacting effects of recruitment, behavior, and physical disturbance. *Ecol. Monogr.* 62:525-46.

Breitburg DL, Loher T, Pacey CA, Gerstein A. 1997. Varying effects of low dissolved oxygen on trophic interactions in an estuarine food web. *Ecol. Monogr.* 67:489-507.

Bricker S, Clement C, Frew S, Harmon M, Harris M, Pirhalla D. 1997. NOAA's Estuarine Eutrophication Survey. Volume 4: Gulf of Mexico Region. Silver Spring, MD. Office of Ocean Resources Conservation Assessment. 46 pp.

Bricker SB, Clement CG, Pirhalla DE, Orlando SP, Farrow DRG. 1999. National estuarine eutrophication assessment. Effects of nutrient enrichment in the Nation's estuaries, NOAA—NOS Special Projects Office, 1999.

Bricker SB, Lipton D, Mason A, Dionne M, Keeley D, Krahforst C, Latimer J, Pennock J. 2006. Improving methods and indicators for evaluating coastal water eutrophication: A pilot study in the Gulf of Maine. NOAA technical report 20. http://ccma.nos.noaa.gov/news/feature/GulfofMaine.html.

Bricker S, Longstaff B, Dennison W, Jones A, Boicourt K, Wicks C, Woerner J. 2007. Effects of nutrient enrichment in the Nation's estuaries: A decade of change. NOAA Coastal Ocean Program Decision Analysis Series No. 26. National Centers for Coastal Ocean Science, *Silver Spring,* MD. 328 p.

Britt NW. 1955. Stratification in western Lake Erie in summer of 1953: Effects on the Hexagenia (Ephemeroptera) population. Ecology 36:239-244.

Brosnan TM, O'Shea ML. 1996. Long-term improvements in water quality due to sewage abatement in the lower Hudson River. *Estuaries* 19:890-900.

Brown CA, Nelson WG, Boese BL, DeWitt TH, Eldrige PM, Kaldy JE, Lee H II, Power JH, Young DR. 2007. An approach to developing nutrient criteria for Pacific northwest estuaries: A case study of Yaquina Estuary, Oregon. USEPA Office of Research and Development, National Health and Environmental Effects Laboratory, Western Ecology Division. EPA/600/R-07/046. 169 pp.

Brunner CA, Beall JM, Bentley SJ, Furukawa Y. 2006. Hypoxia hotspots in the Mississippi Bight. *J. Foram. Res.* 36:95-107.

Burnett LE. 1997. The challenges of living in hypoxic and hypercapnic aquatic environments. *Am. Zool.* 37:633-640.

Buzzards Bay National Estuary Program (NEP). 2008. Accessed June 2008: http://www.buzzardsbay.org/charact.htm.

Buzzelli CP, Luettich RA Jr., Powers SP, Peterson CH, McNinch JE, Pinckney JL, Paerl HW. 2002. Estimating the spatial extent of bottom-water hypoxia and habitat degradation in a shallow estuary. *Mar. Eco. Prog. Ser.* 230:103- 112.

Byrnes MR, Hammer RM, Vittor BA, Ramsey JS, Snyder DB, Bosma KF, Wood JD, Thibaut TD, Phillips NW. 1999. Environmental survey of identified sand resource areas offshore Alabama: Vol. I: Main Text. US Department of Interior, Minerals Management Service, International Activities and Marine Minerals Division, Herndon, VA. OCS Report MMS 99-0052, 326 pp.

Chamberlain R, Hayward D. 1996. Evaluation of water quality and monitoring in the St. Lucie Estuary, Florida. *Water Resor. Bull.* 32:681-696.

Chan F, Barth JA, Lubchenco J, Kirincich A, Weeks H, Peterson WT, Menge BA. 2008. Emergence of anoxia in the California current large marine ecosystem. *Science* 319:920.

CBP (Chesapeake Bay Program). 2008. Water quality database. Accessed December 2008. http://www. chesapeakebay.net/data_waterquality.aspx.

Christensen J, Packard T. 1976. Oxygen utilization and plankton metabolism in a Washington fjord. *Estuar. Costal Mar. Sci.* 4:339-347.

Cicchetti G, Latimer JS, Rego SA, Nelson WG, Bergen BJ, Coiro LL. 2006. Relationships between near-bottom dissolved oxygen and sediment profile camera measures. *J. Mar. Syst.* 62:124-141.

Cloern JE, Oremland RS. 1983. Chemistry and microbiology of a sewage spill in south San Francisco Bay. *Estuaries* 6:399-406.

Collias EE. 1985. Nationwide review of oxygen depletion and eutrophication in estuarine and coastal waters: Pacific region. Report to US Dept. of Commerce, NOAA, National Ocean Service. Rockville, MD.

Cooper SR, Brush GS. 1991. Long-term history of Chesapeake Bay anoxia. *Science* 254: 992-996.

Craig JK, Crowder LB. 2005. Hypoxia-induced habitat shifts and energetic consequences in Atlantic croaker and brown shrimp on the Gulf of Mexico shelf. *Mar. Eco. Prog. Ser.* 294:79-94.

Cronin TM, Vann CD. 2003. The sedimentary record of climatic and anthropogenic influence on the Patuxent estuary and Chesapeake Bay ecosystems. *Estuaries* 26:196-209.

Dauer DM, Rodi AJ Jr., Ranasinghe JA. 1992. Effects of low dissolved oxygen events on the macrobenthos of the lower Chesapeake Bay. *Estuaries* 15:384-391.

D'Avanzo C, Kremer JN. 1994. Diel oxygen dynamics and anoxic events in an eutrophic estuary of Waquoit Bay, Massachusetts. Estuaries 17:131-139.

Deacutis CF, Murray D, Prell W, Saarman E, Korhun L. 2006. Hypoxia in the upper half of Narragansett Bay, RI, during August 2001 and 2002. *Northeastern Natural.* 13:173–198.

Delorme LD. 1982. Lake Erie oxygen: the prehistoric record. Can. *J. Fish. Aquatic Sci.* 39:1021-1029.

deWitt P, Daiber FC. 1974. The hydrography of the Murderkill Estuary, Delaware. *Chesapeake Science* 15:84-95.

Diaz RJ, Neubauer RJ, Schaffner LC, Pihl L, Baden SP. 1992. Continuous monitoring of dissolved oxygen in an estuary experiencing periodic hypoxia the effect of hypoxia on macrobenthos and fish. *Sci. Total Environ. Suppl.* 1992:1055-1068.

Diaz RJ, Rhoads DC, Blake JA, Kropp RK, Keay KE. 2008. Long-term trends of benthic habitats related to reduction in wastewater discharge to Boston Harbor. Estuar. *Coasts* 31:1184–1197.

Edwards WJ, Conroy JD, Culver DA. 2005. Hypolimnetic oxygen depletion dynamics in the central basin of Lake Erie. *J. Great Lakes Res.* 31:262-271.

Engle VD, Summers JK, Macauley JM. 1999. Dissolved oxygen conditions in northern Gulf of Mexico estuaries. *Environ. Monitor. Assess.* 57:1-20.

Engle VD, Summers JK. 1999. Refinement, validation, and application of a benthic condition index for northern Gulf of Mexico estuaries. *Estuaries* 22:624–635.

Figley W, Pyle B, Halgren B. 1979. Socioeconomic impacts. In: Oxygen depletion and associated benthic moralities in New York Bight, 1976. R.L. Swanson, C.J. Sindermann (eds.). NOAA Professional Paper 11, pp. 315-322.

Flemer DA, Kruczynski WL, Ruth BF, Bundrick CM. 1999. The relative influence of hypoxia, anoxia, and associated environmental factors as determinants of macrobenthic community structure in a northern Gulf of Mexico. *J. Aquatic Ecosys. Stress Recov.* 6:311-328.

Frankenberg D. 1976. Oxygen in a tidal river: Low tide concentrations correlates linearly with location. Estuar. *Coast. Mar. Sci.* 4:455-460.

Fritz C, LaBrecque E, Tober J, Behr PJ, Valiela I. 1996. Shrimp in Waquoit Bay: Effects of nitrogen loading on size and abundance. General Scientific Meetings of the Marine Biological Laboratory, Woods Hole, MA 191:326-327.

Garlo EV, Milstein CB, Jahn AE. 1979. Impact of hypoxic conditions in the vicinity of Little Egg Inlet, New Jersey in summer 1976. Estuar. *Coast. Mar. Sci.* 8:421-432.

Gaston G. 1985. Effects of hypoxia on macrobenthos of the inner shelf off Cameron, Louisiana. Estuar. *Coast. Shelf Sci.* 20:603-613.

Gerlofsma J, Ciborowski JJH. 1999. Distribution, density, and influence of hypoxia on eggs of Hexagenia mayflies (Ephemeroptera: Ephemeridae) in western Lake Erie. IAGLR '99. International Association for Great Lakes Research: Great Lakes, Great Science, Great Cities. *Program and Abstracts.* p. A-36.

Gillett DJ, Holland AF, Sanger DM. 2005. Secondary production of a dominant oligochaete (Monopylephorus rubroniveus) in the tidal creeks of South Carolina and its relation to ecosystem characteristics. *Limnol. Oceanogr.* 50:566-577.

Glenn S, Crowley M, Haidvogel D, Song Y. 1996. Underwater observatory captures coastal upwelling event off New Jersey. *Eos, Trans. Am. Geophy.* Union 77:223-236.

Glenn S, Arnone R, Bergmann T, Bissett WP, Crowley M, Cullen J, Gryzmski J, Haidvogel D, Kohut J, Moline M, Oliver M, Orrico C, Sherrell R, Song T, Weidemann A, Chant T, Schofield O. 2004. Biogeochemical impact of summertime coastal upwelling on the New Jersey shelf. *J. Geophy. Res.* 109:C12S02.

Grantham BA, Chan F, Nielsen KJ, Fox DS, Barth JA, Huyer A, Lubchenco J, Menge BA. 2004. Upwelling-driven nearshore hypoxia signals ecosystem and oceanographic changes in the northeast Pacific. Nature 429:749-754.

Gunter G. 1942. Offatts Bayou, a locality with recurrent summer mortality of marine organisms. *Am. Midl. Natural.* 28:631-633.

Gunter G, Williams RH, Davis CC, Smith FGW. 1948. Catastrophic mass mortality of marine animals and coincident phytoplankton bloom on the west coast of Florida, November 1946 to August 1947. *Ecol. Monogr.* 18:309-324.

Haas LW. 1977. The effect of the spring-neap tidal cycle on the vertical salinity structure of the James, York and Rappahannock River, Virginia. USA. *Estuar. Coast. Mar. Sci.* 5:485-496.

Hagy JD, Murrell MC. 2007. Susceptibility of a northern Gulf of Mexico estuary to hypoxia: an analysis using box models. Estuar. *Coast. Shelf Sci.* 74:239-253.

Hagy JD, Boynton WR, Keefe CW, Wood KV. 2004. Hypoxia in Chesapeake Bay, 1950-2001: Long-term change in relation to nutrient loading and river flow. *Estuaries* 27: 634-658.

Hagy JD III, Lehrter JC, Murrell MC. 2006. Effects of Hurricane Ivan on water quality in Pensacola Bay, Florida. *Estuar. Coasts* 29:919–925.

Harper DE Jr., Rabalais NN. 1995. Responses of benthonic and nektonic organisms, and communities, to severe hypoxia on the inner continental shelf of Louisiana and Texas. In: Proceedings of the First Gulf of Mexico Hypoxia Management Conference, Dec 5-6 1995, Kenner, Louisiana, USEPA-55-R-97-002. pp. 41-56.

Harper DE, McKinney LD, Salzer RB, Case RJ. 1981. The occurrence of hypoxic bottom water off the upper Texas coast and its effects on the benthic biota. *Contrib. Mar. Sci.* 24:53-79.

Harper DE, McKinney LD, Nance JM, Salzer RB. 1991. Recovery responses of two benthic assemblages following an acute hypoxic event on the Texas continental shelf, northwestern Gulf of Mexico. In: Modern and ancient continental shelf anoxia. R.V. Tyson and T.H. Pearson (eds.). Geological Society Special Publication No. 58, London. pp. 49-64.

Hartman WL. 1972. Lake Erie: effects of exploitation, environmental changes, and new species on the fishery resources. *J. Fish. Res. Board Can.* 29:931-936.

Heck KL. 1976. Community structure and the effects of pollution in sea grass meadows and adjacent habitats. *Mar. Biol.* 35:345-357.

Hobbie JE, Copeland BJ, Harrison WG. 1975. Sources and fates of nutrients in the Pamlico River estuary of North Carolina. In: L. E. Cronin (ed.), Estuarine Research. Vol. 1. Academic Press, New York. pp. 287-302.

Holland AF, Shaughnessy AT, Hiegel MH. 1987. Long-term variation in mesohaline Chesapeake Bay macrobenthos: spatial and temporal patterns. *Estuaries* 10:370-378.

Howell P, Simpson D. 1994. Abundance of marine resources in relation to dissolved oxygen in Long Island Sound. *Estuaries* 17:394-402.

Jassby A, Van Nieuwenhuyse EE. 2005. Low dissolved oxygen in an estuarine channel (San Joaquin River, California): mechanisms and models based on long-term time series. San Francisco Estuary Watershed Sci. Vol. 3.

Justic′ D, Rabalais NN, Turner RE. 1996. Effects of climate change on hypoxia in coastal waters: a doubled CO2 scenario for the northern Golf of Mexico. *Limnol. Oceanogr.* 41:992-1003.

Karna DW. 2003. A review of some of the effects of reduced dissolved oxygen on the fish and invertebrate resources of Ward Cove, Alaska. Report to Watershed Restoration Unit, Office of Water, US Environmental Protection Agency, Region 10, Seattle, WA. 32 p.

Keefer ML, Peery CA, Wright N, Daigle WR, Caudill CC, Clabough TS, Griffith DW, ZachariasMA. 2008. Evaluating the NOAA Coastal and Marine Ecological Classification Standard in estuarine systems: A Columbia River Estuary case study. *Estuarine, Coastal and Shelf Science* 78: 89-106.

Keister JE, Houde ED, Breitburg DL. 2000. Effects of bottom-layer hypoxia on abundances and depth distributions of organisms in Patuxent River, Chesapeake Bay. Mar. *Ecol. Prog. Ser.* 205:43-59.

Kemp WM, Boynton WR, Adolf JE, Boesch DF, Boicourt WC, Brush G, Cornwell JC, Fisher TR, Glibert PM, Hagy JD, Harding LW, Houde ED, Kimmel DG, Miller WD, Newell RIE, Roman MR, Smith EM, Stevenson JC. 2005.

Eutrophication of Chesapeake Bay: historical trends and ecological interactions. Mar. *Ecol. Prog. Ser.* 303:1-29.

Koepfler E, Lake S, Smith EM, Bennett J, Libes S. 2007. Examination of inner shelf water quality in Long Bay, South Carolina, using dataflow. Estuarine Research Federation Providence Rhode Island Meeting, November 2007, abstract.

Krieger KA. 1985. Snail distributions in Lake Erie: the influence of anoxia in the southern central basin nearshore zone. *Ohio J. Sci.* 85:230-244.

Krieger KA, Bur MT, Ciborowski JJH, Barton DR, Schloesser DW. 2007. Distribution and abundance of burrowing mayflies (Hexagenia spp.) in Lake Erie, 1997–2005. *J. Great Lakes Res.* 33:20-33.

Kuo AY, Park K, Moustafa MZ. 1991. Spatial and temporal variabilities of hypoxia in the Rappahannock River, Virginia. *Estuaries* 14:113-121.

Landsberg JH, Flewelling LJ, Naar J. 2009. Karenia brevis red tides, brevetoxins in the food web, and impacts on natural resources: Decadal advancements. *Harmful Algae* 8(4): 598-607.

Lapointe BE, Matzie WR. 1996. Effects of stormwater nutrient discharges on eutrophication processes in nearshore waters of the Florida Keys. *Estuaries* 19:422-435.

Lee HW, Bailey-Brock JH, McGurr MM. 2006. Temporal changes in the polychaete infaunal community surrounding a Hawaiian mariculture operation. *Mar. Ecol. Prog. Ser.* 307:175-185.

Lee YJ, Lwiza KMM. 2007. Characteristics of bottom dissolved oxygen in Long Island Sound, New York. *Estuar. Coast. Shelf Sci.* 76:187-200.

Lehman PW, Sevier J, Giulianotti J, Johnson M. 2004. Sources of oxygen demand in the lower San Joaquin River, California. *Estuaries* 27:405-418.

Lenihan HS. 1999. Physical-biological coupling on oyster reefs: how habitat structure influences individual performance. Ecol. Monogr. 69:251-275.

Lenihan HS, Peterson CH. 1998. How habitat degradation through fishery disturbance enhances impacts of hypoxia on oyster reefs. *Ecol. Appl.* 8:128-140.

Leonard CL, McClintock JB. 1999. The population dynamics of the brittlestar Ophioderma brevispinum in near- and farshore seagrass habitats of Port Saint Joseph Bay, Florida. *Gulf of Mexico Science* 17:87-94.

Lerberg SB, Holland AF, Sanger DM. 2000. Responses of tidal creek macrobenthic communities to the effects of watershed development. *Estuaries* 23:838-853.

Leverone JR. 1995. Diurnal dissolved oxygen in two Tampa Bay seagrass meadows: ramifications for the survival of adult bay scallops (Argopectin irradians concentricus). *Florida Scient.* 58:141-152.

Livingston RJ. 1975. Impact of kraft pulp-mill effluents on estuarine and coastal fishes in Apalachee Bay, Florida, USA. *Mar. Biol.* 32:19-48.

Llansó RJ. 1992. Effects of hypoxia on estuarine benthos: the lower Rappahannock River (Chesapeake Bay), a case study. *Estuar. Coast. Shelf Sci.* 35:491-515.

Loesch H. 1960. Sporadic mass shoreward migrations of demersal fish and crustaceans in Mobile Bay, Alabama. Ecology 41:292–98.

Lowery TA. 1998. Modeling estuarine eutrophication in the context of hypoxia, nitrogen loadings, stratification and nutrient ratios. *J. Environ. Manag.* 52:289-305.

Luther GW, Ma SF, Trouwborst R, Glazer B, Blickley M, Scarborough RW, Mensinger MG. 2004. The roles of anoxia, H2S, and storm events in fish kills of dead-end canals of Delaware inland bays. *Estuaries* 27:551-560.

Maciolek NJ, Diaz RJ, Dahlen DT, Hunt CD, Williams IP. 2005. 2003 Boston Harbor Benthic Monitoring Report. Boston: Massachusetts Water Resources Authority. *Report ENQUAD* 2005-06. 83 p.

MacPherson TA, Cahoon LB, Mallin MA. 2007. Water column oxygen demand and sediment oxygen flux: patterns of oxygen depletion in tidal creeks. *Hydrobiologia* 586:235-248.

Mallin MA, Posey MH, Shank GC, McIver MR, Ensign SH, Alphin TD. 1999. Hurricane effects on water quality and benthos in the Cape Fear watershed: Natural and anthropogenic impacts. *Ecol. Appl.* 9:350-362.

Mallin MA, Johnson VL, Ensign SH, MacPherson TA. 2006. Factors contributing to hypoxia in rivers, lakes, and streams. *Limnol. Oceanogr.* 51:690–701.

Mason WT, Jr. 1998. Macrobenthic monitoring in the Lower St. Johns River, Florida. *Environ. Monitor. Assess.* 50:101-130.

Maxted JR, Eskin RA, Weisberg SB, Chaillou JC, Kutz FW. 1997. The ecological condition of dead-end canals of the Delaware and Maryland coastal bays. *Estuaries* 20:319-327.

May E. 1973. Extensive oxygen depletion in Mobile Bay, Alabama. Limnol. *Oceanogr.* 18:353-366.

Melrose DC, Oviatt CA, Berman MS. 2007. Hypoxic events in Narragansett Bay, Rhode Island, during the summer of 2001. *Estuar. Coasts* 30:47-53.

Mobile Bay National Estuary Program (NEP). 2008. Accessed June 2008 http://www.mymobilebay.com/

Montagna PA, Ritter C. 2006. Direct and indirect effects of hypoxia on benthos in Corpus Christi Bay, Texas, USA. *J. Exper. Mar. Biol. Ecol.* 330:119-131.

Montagna PA, Kalke RD. 1992. The effect of freshwater inflow on meiofaunal and macrofaunal populations in the Guadalupe and Nueces Estuaries, Texas. *Estuaries* 15:307-326.

Moser FC. 1998. Sources and sinks of nitrogen and trace metals, and benthic macrofauna assemblages in Barnegat Bay, New Jersey. Dissertation Abstracts International Part B: Science and Engineering, University Microfilms International 58:5849.

National Oceanic and Atmospheric Administration (NOAA). 2008. Case: New Bedford Harbor, MA. Accessed July 2008: http://www.darrp.noaa. gov/northeast/new_bedford/admin.html.

Neponset River Watershed Association (NERWA). 2004. Boston Harbor south watersheds 2004 assessment report. Massachusetts Executive Office of Environmental Affairs, Boston. 180 p.

Newcombe CL, Horne WA. 1938. Oxygen-poor waters of the Chesapeake Bay. *Science* 88:80-81.

Nichols FH, Cloern JE, Luoma SN, Peterson DH. 1986. The modification of an estuary. *Science* 231:567-573.

Officer CB, Biggs RB, Taft JL, Cronin LE, Tyler MA, Boynton WR. 1984. Chesapeake Bay anoxia: origin, development, and significance. *Science* 223:22-27.

Okey TA. 2003. Macrobenthic colonist guilds and renegades in Monterey canyon (USA) drift algae: partitioning multidimensions. *Ecol. Monogr.* 73:415-440.

Osterman LE, Poore RZ, Swarzenski PW. 2007. The last 1000 years of natural and anthropogenic low-oxygen bottom water on the Louisiana shelf, Gulf of Mexico. *Mar. Micropaleo.* 66:291-303.

Paerl HW, Bales JD, Ausley LW, Buzzelli CP, Crowder LB, Eby LA, Go M, Peierls BL, Richardson TL, Ramus JS. 2000. Hurricane's hydrological, ecological effects linger in major US estuary. Eos, *Trans. Am. Geophys. Union* 81:457-462.

Paerl HW, Pinckney JL, Fear JM, Peierls BL. 1998. Ecosystem responses to internal and watershed organic matter loading: Consequences for hypoxia in the eutrophying Neuse River Estuary, North Carolina, USA. *Mar. Ecol.* Prog. Ser. 166:17-25.

Paerl HW, Pinkney JL, Kucera SA. 1995. Clarification of the structural and functional roles of heterocysts and anoxic microzones in the control of pelagic nitrogen fixation. *Limnol. Oceanogr.* 40:634-638.

Parker CA, O'Reilly JE. 1991. Oxygen depletion in Long Island Sound: A historical perspective. Estuaries 14:248-264.

Parker-Stetter SL, Horne JK. 2008. Nekton distribution and midwater hypoxia: A seasonal, diel prey refuge? *Estuar. Coast. Shelf Sci.* 81:13-18.

Patrick R. 1988. Changes in the chemical and biological characteristics of the Upper Delaware River estuary in response to environmental laws. (E. Majumdar, W. Miller, and L. E. Sage, Eds.), pp. 332-359. Pennsylvania Academy of Science, Philadelphia, PA.

Paulson AJ, Curl HC Jr., Freely RA. 1993. The biogeochemistry of nutrients and trace metals in Hood Canal, a Puget Sound fjord. *Mar. Chem.* 43:157-173.

Penland S, Beall A, Waters J, Kindinger J (eds.). 2002. Environmental Atlas of the Lake Pontchartrain Basin. Lake Pontchartrain Basin Foundation, New Orleans, LA. Accessed June 2008: http://pubs.usgs.gov/of/2002/of02-206/.

Pennock JR, Sharp JH, Schroeder WW. 1994. What controls the expression of estuarine eutrophication? Case studies of nutrient enrichment in the Delaware Bay and Mobile Bay estuaries, USA. In: Changes in Fluxes in Estuaries: Implications from Science to Management. Edited by K.R. Dyer, R.J. Orth. Olsen and Olsen, Fredensborg. p. 139-146.

Perez-Domingues et al 2006. Environmental variability in seagrass meadows: effects of nursery environment cycles on growth and survival in larval red drum Sciaenops ocellatus. *Mar. Ecol. Prog. Ser.* 321:41-53.

Pihl L, Baden SP, Diaz RJ, Schaffner LC. 1992. Hypoxia-induces structural changes in the diet of bottom-feeding fish and crustacea. *Mar. Biol.* 112:349-361.

Pihl L, Baden SP, Diaz RJ. 1991. Effects of periodic hypoxia on distribution of demersal fish and crustaceans. *Mar. Biol.* 108:349-360.

Portnoy JW. 1991. Summer oxygen depletion in a diked New England estuary. *Estuaries* 14:122-129.

Posey MH, Alphin TD, Cahoon L, Lindquist D, Becker ME. 1999. Interactive effects of nutrient additions and predation on infaunal communities. *Estuaries* 22:785-792.

Powers SP, Peterson CH, Christian RR, Sullivan E, Powers MJ, Bishop MJ, Buzzelli CP. 2005. Effects of eutrophication on bottom habitat and prey resources of demersal fishes. *Mar. Ecol. Prog. Ser.* 302:233-243.

Rabalais NN. 1998. Oxygen depletion in coastal waters. National Oceanic and Atmospheric Administration's State of the Coast Report. *Silver Spring,* MD: NOAA. 52 p.

Rabalais NN, Turner RE. (eds.) 2001. Coastal hypoxia. Consequences for living resources and ecosystems, Vol. American Geophysical Union, Washington DC.

Rabalais NN, Dagg MJ, Boesch DF. 1985. Nationwide review of oxygen depletion and eutrophication in estuarine and coastal waters: Gulf of Mexico, Appendix 3C. In: Whitledge, T.E. (ed.). Nationwide review of oxygen depletion and eutrophication in estuarine and coastal waters. Report to US Dept. of Commerce, National Ocean Service, Washington DC.

Rabalais NN, Turner RE, Sen Gupta BK, Boesch DF, Chapman P, Murrell MC. 2007. Hypoxia in the Northern Gulf of Mexico: Does the Science Support the Plan to Reduce, Mitigate, and Control Hypoxia? *Estuar. Coasts* 30:753- 772.

Reish DJ. 1955. The relation of polychaetous annelids to harbor pollution. *Public Health Report* 70:1168-1174.

Reish DJ. 2000. The seasonal settlement of polychaete larvae before and after pollution abatement in Los Angeles-Long Beach Harbors, California. *Bull. Mar. Sci.* 67:672.

Rhoads JM, Yozzo DJ, Cianciola MM, Will RJ. 2001. Norton Basin/Little Bay restoration project: historical and environmental background report. US Army Corps of Engineers, New York District, CENAN-PL-ES, New York. p. 50.

Riedel GF, Sanders JG, Osman RW. 1999. Biogeochemical control on the flux of trace elements from estuarine sediments: effects of seasonal and short-term hypoxia. *Mar. Environ. Res.* 47:349-372.

Rikard FS, Wallace RK, Rouse D, Saoud I. 2000. The effect of low oxygen on oyster survival during reef restoration efforts in Bon Secour Bay, Alabama. *J. Shellfish Res.* 19:640.

Ritter C, Montagna PA. 1999. Seasonal hypoxia and models of benthic response in a Texas bay. Estuaries 22:7-20.

Rosa F, Burns NM. 1987. Lake Erie central basin oxygen depletion changes from 1929-1980 *J. Great Lakes Res.* 13:684-696.

Sagasti A, Duffy JE, Schaffner LC. 2003. Estuarine epifauna recruit despite periodic hypoxia stress. *Mar. Biol.* 142:111-122.

Sagasti A, Schaffner LC, Duffy JE. 2001. Effects of periodic hypoxia on mortality, feeding and predation in an estuarine epifaunal community. *J. Exp. Mar. Biol. Ecol.* 258:257-283.

Sale JW, Skinner WW. 1917. The vertical distribution of dissolved oxygen and the precipitation by salt water in certain tidal areas. *J. Franklin Institute* 184:837-848.

Sanger DM, Arendt MD, Chen Y, Wenner EL, Holland AF, Edwards D, Caffrey J. 2002. A synthesis of water quality data: National Estuarine Research Reserve System-wide monitoring program (1995-2000). National Estuarine Research Reserve Technical Report Series 2002:3. South Carolina Department of Natural Resources, Marine Resources Division Contribution No. 500. 135 p.

Santos SL, Simon JL. 1980. Marine soft-bottom community establishment following annual defaunation: larval or adult recruitment. *Mar. Ecol. Prog.* Ser. 2:235-241.

Schimmel SC, Benyi SJ, Strobel CJ. 1999. An assessment of the ecological condition of Long Island Sound, 1990- 1993. *Environ. Monitor. Assess.* 56:27-49.

Segar DA, Berberian GA. 1976. Oxygen depletion in the New York Bight apex: Causes and consequences. American Society of Limnology and Oceanography Special Symposium 2:220-239.

Seitz RD, Marshall LS Jr., Hines AH, Clark KL. 2003. Effects of hypoxia on predator-prey dynamics of the blue crab Callinectes sapidus and the Baltic clam Macoma balthica in Chesapeake Bay. *Mar. Ecol. Prog. Ser.* 257:179-188.

Seliger HH, Boggs JA, Biggley WH. 1985. Catastrophic anoxia in the Chesapeake Bay in 1984. *Science* 228:70–73.

Sen Gupta BK, Turner RE, Rabalais NN. 1996. Seasonal oxygen depletion in the continental shelf waters of Louisiana: Historical record of benthic foraminifers. *Geology* 24:227-230.

Sindermann C, Swanson R. 1980. Chapter 1. Historical and regional perspective. In: Oxygen depletion and associated benthic mortalities in New York Bight, 1976. R.L. Swanson and C.J. Sindermann (eds.) Rockville. US Department of Commerce National Oceanic and Atmospheric Administration. pp.1-16.

Smith RW, Bergen M, Weisberg SB, Cadien D, Dalkey A, Montagne D, Stull JK, Velarde RG. 2001. Benthic response index for assessing infaunal communities on the southern California mainland shelf. *Ecol. Appl.* 11:1073-1087.

Stanley DW. 1985. Nationwide review of oxygen depletion and eutrophication in estuarine and coastal waters: southeast region. Report to US Dept. of Commerce, NOAA, National Ocean Service. Rockville, MD. 354 p.

Stanley DW, Nixon SW. 1992. Stratification and bottom-water hypoxia in the Pamlico River Estuary. *Estuaries* 15:270- 281.

Summers JK, Weisberg CB, Holland AF, Kou J, Engle VD, Breitberg DL, Diaz RJ. 1997. Characterizing dissolved oxygen conditions in estuaries environments. *Environ. Monitor. Assess.* 45:319-328.

Swanson RL, Parker CA. 1988. Physical environmental factors contributing to recurring hypoxia in the New York Bight. *Trans. Am. Fish. Soc.* 117:37-47.

Taylor DI. 2000. Harbor sampling update. MWRA Sewerage Division. Accessed July 2008: http://www.mwra.state. ma.us/harbor/html/ wk0901 00.htm.

Tenore KR. 1972. Macrobenthos of the Pamlico River estuary, North Carolina. Ecol. Monogr. 42:51-69.

TIEE (Teaching Issues and Experiments in Ecology TIEE). 2008. What's killing the coral reefs and seagrasses? Accessed December 2008: http://tiee.ecoed.net/vol/v1/ figure_sets/coral/coral_back1.html.

Tomasko DA, Anastasiou C, Kovach C. 2006. Dissolved oxygen dynamics in Charlotte Harbor and its contributing watershed, in response to hurricanes Charley, Frances, and Jeanne - impacts and recovery. *Estuar. Coasts* 29:932-938.

Turner RE, Rabalais NN. 1994. Coastal eutrophication near the Mississippi river delta. *Nature* 368:619–621.

Turner RE, Rabalais NN, Fry B, Atilla N, Milan CS, Lee JM, Normandeau C, Oswald TA, Swenson EM, Tomasko DA. 2006. Paleo-indicators and water quality change in the Charlotte Harbor estuary (Florida). *Limnol. Oceanogr.* 51:518–533.

Tyler RM. 2004. Distribution and avoidance patterns of juvenile summer flounder (Paralichthys dentatus) and weakfish (Cynoscion regalis) in relation to hypoxia: field studies in a temperate coastal lagoon tributary and laboratory choice-trial experiments. PhD thesis, University of Delaware, Newark, DE.

Tyler RM, Targett TE. 2007. Juvenile weakfish Cynoscion regalis distribution in relation to diel-cycling dissolved oxygen in an estuarine tributary. *Mar. Ecol. Prog. Ser.* 333:257-269.

Verity PG, Alber M, Bricker SB. 2006. Development of hypoxia in well-mixed subtropical estuaries in the southeastern USA. *Estuar. Coasts* 29:665-673.

Wang T. 2005. Hypoxia in shallow coastal waters: a case study in Onancock Creek, Virginia. Masters of Science Thesis, Virginia Institute of Marine Science, College of William and Mary, Gloucester Pt. 129 p.

Washington State Department of Ecology. 2002. Washington State marine water quality, 1998 through 2000. Dept. Ecol. Pub. No. 02-03-056. 111 p.

Weisberg SB, Wilson HT, Himchak P, Baum T, Allen R. 1996. Temporal trends in abundance of fish in the tidal Delaware River. *Estuaries* 19:723-729.

Welsh B, Welsh R, DiGiacomo-Cohen M. 1994. Quantifying hypoxia and anoxia in Long Island Sound. In: Changes in fluxes in estuaries: Implications from science to management. K. Dyer, and R. Orth (eds.) Fredensborg: Olsen and Olsen. pp. 131-137.

White WA, Calnan TR, Morton RA, Kimble RS, Littleton TJ, McGowen JH, Nance HS, Schmedes KE. 1984.

Submerged lands of Texas, Galveston-Houston area: sediments, geochemistry, benthic macroinvertebrates, and associated wetlands. Bureau of Economic Geology, University of Texas at Austin, Austin, Texas. 270 p.

Whitledge TE. 1985. Nationwide review of oxygen depletion and eutrophication in estuarine and coastal waters: northeast region. Report to US Dept. of Commerce, NOAA, National Ocean Service. Rockville, MD. 718 p.

Wilkin RT, Barnes HL. 1997. Pyrite formation in an anoxic estuarine basin. *Am. J. Sci.* 297:620-650.

Windsor JG, Jr. 1985. Nationwide review of oxygen depletion and eutrophication in estuarine and coastal waters. Final Rpt. to Brookhaven Nat. Lab. (New York). Florida Inst. Technol., Melbourne.

Zaikowski L, McDonnell KT, Rockwell RF, Rispoli F. 2008. Temporal and spatial variations in water quality on New York south shore estuary tributaries: Carmans, Patchogue, and Swan Rivers. Estuar. Coasts 31:85-100.

Ziegler S, Benner R. 1998. Ecosystem metabolism in a subtropical, seagrass-dominated lagoon. *Mar. Ecol. Prog. Ser.* 173:1-12.

Zimmerman AR, Canuel EA. 2000. A geochemical record of eutrophication and anoxia in Chesapeake Bay sediments: anthropogenic influence on organic matter composition. *Mar. Chem.* 69:117-137.

In: Assessing the Hypoxia Threat in U.S. Coastal Waters
ISBN 978-1-62081-303-4
Editors: William Jacobs and Steve Frederickson

Chapter 2

THIS IS A WRITTEN TESTIMONY OF ROBERT E. MAGNIEN, DIRECTOR, CENTER FOR SPONSORED COASTAL OCEAN RESEARCH, US DEPARTMENT OF COMMERCE. HEARING ON "THE HARMFUL ALGAL BLOOMS AND HYPOXIA RESEARCH AND CONTROL AMENDMENTS ACT OF 2011"*

INTRODUCTION

Good morning Mr. Chairman and members of the Subcommittee. My name is Robert Magnien and I am the Director of the National Oceanic and Atmospheric Administration's (NOAA) Center for Sponsored Coastal Ocean Research (CSCOR). CSCOR, as one Center of the National Centers for Coastal Ocean Science, provides competitive funding for regional-scale, multidisciplinary research on understanding and predicting the impacts of major stressors on coastal ecosystems, communities, and economies in order to support informed, ecosystem-based management. In this capacity, I administer the five national programs solely focused on harmful algal blooms (HAB) and hypoxia that were authorized by the *Harmful Algal Bloom and Hypoxia Research and Control Act of 1998* (*HABHRCA*) and reauthorized in 2004. I serve on the Interagency Working Group on Harmful Algal Blooms, Hypoxia, and Human Health of the Subcommittee on Ocean Science and Technology to coordinate NOAA's programs with other federal agencies. I also serve as NOAA's representative to the Coordinating Committee for the Mississippi River/Gulf of Mexico Watershed Nutrient Task Force, which is addressing the Nation's largest hypoxic zone in the northern Gulf of Mexico. Additionally, I serve as the US representative and Chair of the Intergovernmental Oceanographic Commission panel on HABs to maximize international opportunities for exchange of relevant research and management information. Though I am the Director of

* This is an edited, reformatted and augmented version of written testimony given on June 1, 2011 before the Subcommittee on Energy and Environment, Committee on Science, Space, and Technology, US House of Representatives.

CSCOR, this testimony speaks about programs across NOAA, where multiple offices work together to achieve mission goals.

NOAA's mandate includes protecting the lives and livelihoods of Americans, and providing products and services that benefit the economy, environment, and public safety of the Nation. By improving our understanding of, and ability to predict changes in, the Earth's environment, and by conserving and managing ocean and coastal resources, NOAA generates tremendous value for the Nation. NOAA's role is all the more important given the profound economic, environmental, and societal challenges currently facing the country. Two of these challenges are HABs and hypoxia, which cause significant adverse human health and economic impacts.

HABs, which now occur in all US states,[1,2] are a growing problem worldwide. HABs threaten human and ecosystem health, and the vitality of fish and shellfish, protected species, and coastal economies. Similarly, hypoxia occurs in over 300 US coastal ecosystems,[3] including the Great Lakes. There has been a 30-fold increase in hypoxia events since 1960,[3] signaling severe degradation of water quality and aquatic habitats nation-wide. HABs and hypoxia are two of the most complex phenomena currently challenging management of aquatic ecosystems. Given the profound, pervasive, complex and growing impacts of HABs and hypoxia, these are important issues NOAA will continue to address in the coming years.

At this very time, with unprecedented amounts of freshwater and associated nutrients and other chemicals entering the Gulf of Mexico from the Mississippi River Basin, we are witnessing some of these complex factors that drive HABs and hypoxia. NOAA's longstanding HABHRCA research has demonstrated the relationship between nutrient inputs and hypoxia and provided the ability to forecast the size of the hypoxic zone both in the short-term and for long-term management purposes. In addition, NOAA's spring flood outlook, issued in mid-February, indicated a "high risk" for flooding along the Mississippi. Based on the high flows and expected high nutrient loads which will be measured by the US Geological Survey, this year's zone will likely be one of the largest ever. NOAA will issue its annual forecast for the size of the hypoxic zone in June. There is also the potential for toxic algal blooms to develop in Lake Pontchartrain as has been the case in the past when floodwaters have been diverted into the lake. NOAA is moving on a number of fronts to assist in the response to the flooding, including adding capabilities onto its existing HABHRCA research, monitoring, and response projects in the Gulf region in order to provide local, state, and federal officials with the latest and most scientifically accurate information on these coastal impacts.

I appreciate the opportunity to update the Subcommittee on major accomplishments in NOAA's HAB and hypoxia programs. I will first describe the nature of the problem in more detail, then discuss NOAA's role in addressing HABs and hypoxia in our coastal and Great Lakes waters, and conclude with some of the significant advances that NOAA has made as a result of HABHRCA.

Harmful Algal Blooms in the United States

Algae are simple plants that, in general, are beneficial because they provide the main source of energy that sustains aquatic life. However, some algae cause harm to humans, animals, and the environment by producing toxins or by growing in excessively large numbers. When this occurs they are referred to as "harmful algal blooms" or HABs. Sometimes, certain algal species accumulate in such high numbers that they discolor the water, and are commonly referred to as "red tides" or "brown tides." Figure 1 lists some of the major HAB-causing organisms in the United States.

Some algae produce potent toxins that cause illness or death in humans and other organisms. Fish, seabirds, manatees, sea lions, turtles, and dolphins are some of the animals commonly affected by harmful algae. Humans and other animals can be exposed to algal toxins through the food they eat, the water they drink or swim in, or the air they breathe. Other algae species, although nontoxic to humans and wildlife, form such large blooms that they degrade habitat quality through massive overgrowth, shading, and oxygen depletion (hypoxia), which occurs after the bloom ends and the algae decay. These high biomass blooms can also be a nuisance to humans when masses of algae accumulate along beaches and subsequently decay.

HABs can have major negative impacts on local economies when, for example, shellfish harvesting is restricted to protect human health or when tourism declines due to degradation of recreational resources. HABs can also result in significant public health costs when humans become ill. A recent estimate[4] suggests that HABs occurring in marine waters alone have an average annual impact of $82 million in the United States. This is a conservative estimate since comprehensive data that includes the various economic impacts of all major blooms is not available. In 2005, a single HAB event in New England resulted in a loss of $18 million in shellfish sales in Massachusetts alone.[5] Economic impacts can be difficult to calculate as they vary from region to region and event to event, but they are a primary concern of coastal communities that experience HAB events.

In addition to impacting public health, ecosystems, and local economies, HABs can also have significant social and cultural consequences. For example, along the Washington and Oregon coasts, tens of thousands of people visit annually to participate in the recreational harvest of razor clams. However, a series of beach closures in recent years due to high levels of the HAB toxin domoic acid prevented access to this recreational fishery. These harvesting closures have not only caused economic losses, they have also resulted in an erosion of community identity, community recreation, and a traditional way of living for native coastal cultures.

As mentioned above, the geographic distribution of HAB events in the United States is broad. All coastal states have experienced HAB events in marine waters in the last decade, and freshwater HABs occur in the Great Lakes and in many inland waters. Evidence indicates the frequency and distribution of HAB events and their associated impacts have increased considerably in recent years in the United States and globally.[6]

Although all coastal states experience HABs, the specific organisms responsible for the HABs differ among regions of the country (see Figure 1). As a result, the harmful impacts experienced vary in their type, scope and severity, which led to the need for specific management approaches for each region and species and region-specific scientific understanding to support an effective and efficient management response. Some species need to be present in very high abundances before harmful effects occur, which makes it easier to detect and track the HAB. However, other species cause problems at very low concentrations, essentially being hidden among other benign algae, making them difficult to detect and track. The factors that cause and control HABs, from their initiation to their decline vary, not only by species, but also by region due to differences in local factors such as the shape of the coastline, runoff patterns, oceanography, nutrient regime, other organisms present in the water, etc. Consequently, the development of HAB management strategies requires a regional approach.

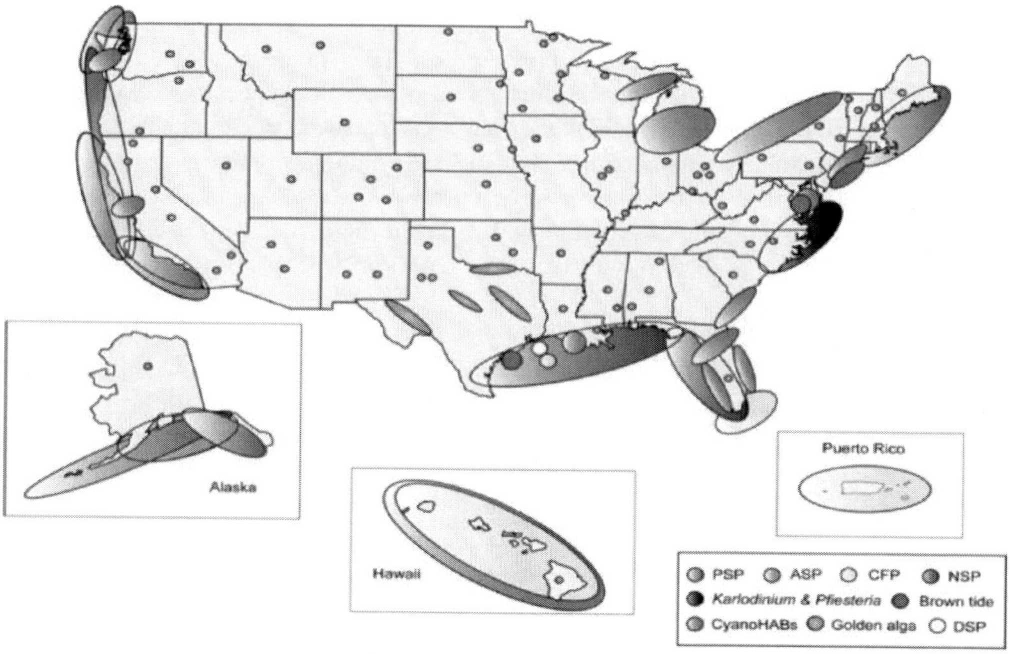

Figure 1. Distribution of major HAB syndromes and events in the US Source: NOAA U. S. National Office for Harmful Algal Blooms (http://www.whoi.edu/redtide/page. do?pid=14898), Woods Hole, MA.

As noted above, the causes of HABs are complex and are controlled by a variety of factors. While we know that the underlying causes leading to HAB development vary between species and locations, we do not have a full understanding of all the factors involved, including the interplay of different contributing factors. In general, algal species grow best when environmental conditions (such as temperature, salinity, and availability of nutrients and light) are optimal for cell growth. Other biological and physical processes (such as predation, disease, toxins and water currents) determine whether enhanced cell growth will result in biomass accumulation (or what we call a "bloom"). The challenge for understanding the causes of HABs stems from the complexity and interrelationship of these processes for individual species and among different HAB species. The complexity of interactions between HABs, the environment, and other plankton further complicates the predictions of when and where HAB events will occur. Knowledge of how these factors control the initiation, sustainment, and decline of a bloom is a critical precursor for advancing HAB management.

Human activities are thought to contribute to the increased frequency of some HABs. For example, increased nutrient pollution has been acknowledged as a factor contributing to increased occurrence of several high biomass HABs.[7] Other human-induced environmental changes that may foster development of certain HABs include changes in the types of nutrients entering coastal waters, alteration of food webs by overfishing, introductions of non-indigenous species that change food web structure, introduction of HAB cells to new areas via ballast water or other mechanisms, and modifications to water flow. It should also be noted that climate change will almost certainly influence HAB dynamics in some way since many critical processes governing HAB dynamics — such as temperature, water column stratification, upwelling and ocean circulation patterns, and freshwater and land-derived nutrient inputs — are influenced by climate. The interactive role of climate change with the other factors driving the frequency and severity of HABs is in the early stages of research, but climate change is expected to exacerbate the HAB problem in some regions and shift species distributions geographically. (http://www.cop.noaa.gov/stressors/extremeevents/hab/current/CC_habs.aspx).

Hypoxia in the US

Hypoxia means "low oxygen." In aquatic systems, low oxygen generally refers to a dissolved oxygen concentration less than 2 to 3 milligrams of oxygen per liter of water (mg/L), but sensitive organisms can be affected at higher thresholds (e.g. 4.5 mg/L). A complete lack of oxygen is called anoxia. Hypoxic waters generally do not have enough oxygen to support fish and other aquatic animals, and are sometimes called dead zones because the only organisms that can live there are tolerant microbes.

The incidence of hypoxia has increased 10-fold globally in the past 50 years and almost 30-fold in the US since 1960, with over 300 coastal ecosystems[3] now experiencing hypoxia (see Figure 2). The increasing occurrence of hypoxia in coastal waters represents a significant threat to the health and economy of our Nation's coasts and Great Lakes. This trend is exemplified most dramatically off the coast of Louisiana and Texas, where the second largest eutrophication-related hypoxic zone in the world is associated with the nutrient pollutant load discharged by the Mississippi and Atchafalaya Rivers.

Although coastal hypoxia can be caused by natural processes, the dramatic increase in the incidence of hypoxia in US waters is linked to eutrophication due to nutrient (nitrogen and phosphorus) and organic matter enrichment, which has been accelerated by human activities. Sources of enrichment include point source discharges of wastewater, nonpoint source atmospheric deposition, and nonpoint source runoff from croplands, lands used for animal agriculture, and urban and suburban areas.

The difficulty of reducing nutrient inputs to coastal waters results from the close association between nutrient loading and a broad array of human activities in watersheds and explains the growth in the number and size of hypoxic zones. While nutrients leaving water treatment facilities can often be controlled through improvements in technology and facility upgrades, diffuse runoff from nonpoint sources, such as agriculture, is more difficult to control. Conservation programs, such as those administered by USDA's Natural Resources Conservation Service, play an important part in helping to reduce edge of field runoff from agricultural operations. Although progress has been made in recent years to better optimize nutrient application through the development of nutrient management plans and best practices, agriculture remains a leading source of nutrient pollution in many watersheds due in part to the high demand for nitrogen intensive crops. Another exacerbating factor is the short-circuiting of water flow due to drainage practices, including tile drainage and ditching, that have been used to convert wetlands to croplands. The USDA Agricultural Research Service has led recent efforts to design drainage control structures to increase retention time and denitrification in drainage systems. Wetlands serve as filters and can reduce the transport of nitrogen and phosphorus into local waterways and ultimately coastal waters. Atmospheric nitrogen deposition from fossil fuel combustion remains an important source of diffuse nutrient pollution for rivers and coastal waters.

Unfortunately, hypoxia is not the only stressor impacting coastal ecosystems. Overfishing, HABs, toxic contaminants, and physical alteration of coastal habitats associated with coastal development are several problems that co-occur with hypoxia and interact to decrease the ecological health of coastal waters and reduce the ecological services they can provide.

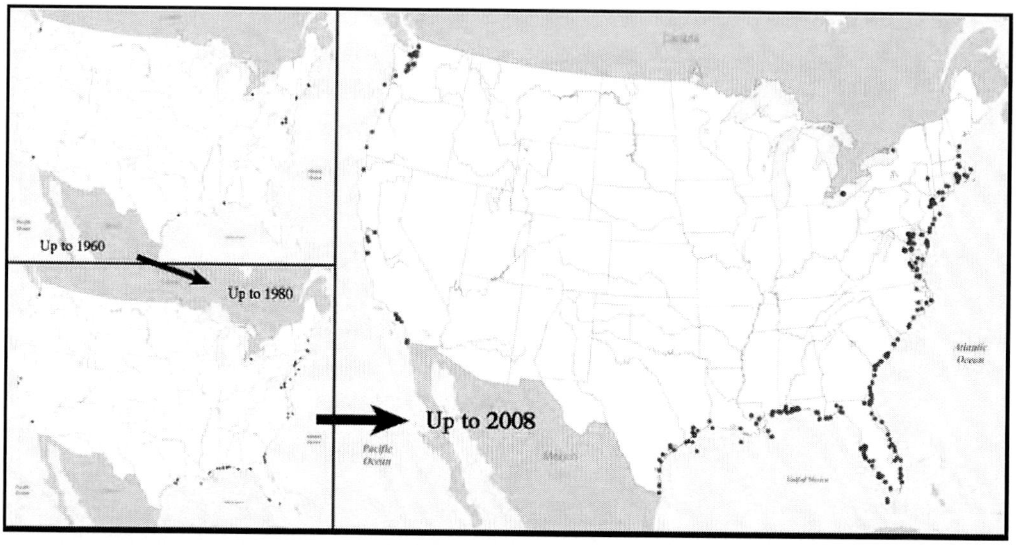

Figure 2. Change in number of US coastal areas experiencing hypoxia from 12 documented areas in 1960 to over 300 now. Not shown here are one hypoxic system in Alaska and one in Hawaii. Source: adapted from Committee on Environment and Natural Resources. 2010[3]

HABHRCA Today

HABHRCA authorizes NOAA to take action to address the growing problem of HABs and hypoxia in the United States. The existing statute:

1. Establishes a mechanism for interagency coordination through an interagency Task Force;
2. Requires reports assessing the causes and impacts of HABs and hypoxia and plans to improve management and response; and
3. Authorizes funding for HAB and hypoxia research through national competitive research programs, and for research and assessment within NOAA.

Since 2005, the Interagency Working Group on HABs, Hypoxia and Human Health within the National Ocean Council has been meeting regularly to coordinate interagency efforts with regard to HABs and hypoxia. A major focus for the group has been developing the five reports mandated by the 2004 reauthorization of HABHRCA. The reports were submitted to Congress between 2007 and 2010 (http://www.cop.noaa.gov/stressors/ extremeevents/ hab/habhrca/Report_Plans.aspx).

These reports provide guidance for NOAA HAB and hypoxia programs as well as other federal or state-supported research that may address aspects of these topics. Specifically, the *HAB Management and Response: Assessment and Plan*[8] recommended the formation of the Prevention, Control, and Mitigation of HABs Program, which NOAA established in 2009. The Plan also highlights the need for an enhanced HAB event response program and a new infrastructure program, which were incorporated into legislation to reauthorize HABHRCA in the 111th Congress.

NOAA HAB and Hypoxia Programs

The goal of NOAA's programs is to prevent or reduce the occurrence of HABs and hypoxia and/or to minimize their impacts. Developing useful products for HAB and hypoxia management is a multi-step process that requires a variety of approaches, and must be based on a strong scientific understanding of the causes and impacts of HABs and hypoxia.

NOAA leads the Nation's three competitive research programs solely focused on HABs and authorized by HABHRCA:

1. The Ecology and Oceanography of Harmful Algal Blooms (ECOHAB) Program is focused on research to determine the causes and impacts of HABs. The ECOHAB Program provides information and tools necessary for developing technologies for, and approaches to, predicting, preventing, monitoring and controlling HABs.
2. The Monitoring and Event Response for Harmful Algal Blooms (MERHAB) Program focuses on incorporating tools, approaches, and technologies from HAB research programs into existing HAB monitoring programs. MERHAB also establishes partnerships to enhance existing, and initiate new, HAB monitoring capabilities to provide managers with timely information needed to mitigate HAB impacts on coastal communities.
3. The newer Prevention, Control, and Mitigation of HABs (PCM HAB) Program, transitions promising prevention, control, and mitigation technologies and strategies to end users. The PCM HAB Program also assesses the social and economic costs of HAB events, and strategies to prevent, control and mitigate those events, which will aid managers in devising the most cost-effective management approaches.

HABHRCA also authorizes research on hypoxia to assess the causes and impacts of this serious problem in order to guide scientifically sound management programs to reduce hypoxic zones and thereby protect valuable marine resources, their habitats and coastal economies. NOAA leads the Nation's two competitive research programs solely focused on hypoxia and authorized by HABHRCA.

1. The Northern Gulf of Mexico Ecosystem and Hypoxia Assessment Program (NGOMEX) supports multiyear, interdisciplinary research projects to inform

management in ecosystems affected by Mississippi/Atchafalaya River inputs. NGOMEX supports research with a focus on understanding the causes and effects of the hypoxic zone over the Louisiana-Texas-Mississippi continental shelf and the prediction of hypoxia's future extent and impacts.

2. The Coastal Hypoxia Research Program (CHRP) supports multiyear, interdisciplinary research projects to inform management of hypoxic zones in all of the Nation's coastal waters except those covered by NGOMEX. The objective of CHRP is to provide research results and modeling tools, which will be used by coastal resource managers to assess alternative management strategies for preventing or mitigating the impacts of hypoxia on coastal ecosystems, and to make informed decisions regarding this important environmental stressor.

HABHRCA authorizes NOAA to carry out research and assessment activities, which has led to a world-class intramural research program on HABs. Much of this research is conducted by scientists from the National Centers for Coastal Ocean Science in collaboration with external partners, including academic researchers, state and federal resource and public health managers, and private enterprises. Active areas of research include HAB and hypoxia forecasting, development of new methods of HAB cell and toxin detection, and understanding the impacts of HAB toxins on higher levels in the food web, including fish, mammals, and humans.

NOAA's extramural and intramural research is leading to the development of a number of operational activities that provide valuable products and assistance. For example, NOAA currently provides HAB forecasts for Florida and Texas coastal waters (http://tidesand currents.noaa.gov/hab/) and has developed plans for a National HAB Forecasting System, which will make routine forecasts in any areas where HABs are a major threat. Forecasts in the Great Lakes, the Gulf of Maine, and the Pacific Northwest are in various stages of development and transition to operations through a combination of extra- and intramural research efforts.

NOAA scientists have been instrumental in developing citizen HAB monitoring networks around the country. Additionally, the NOAA Analytical Response Team (http://www.chbr. noaa.gov/habar/eroart.aspx) and the Wildlife and Algal Toxin Research and Response Network (WARRN-West, http://www.nwfsc.noaa.gov/warrnwest/) provide state-of-the-art toxin analyses during HAB events, especially events that involve unusual animal mortality.

NOAA coordinates and collaborates across the agency on HABHRCA-authorized HAB and hypoxia programs and related efforts to address high priority needs for research, observations, and forecasting.

Many of NOAA's HAB and hypoxia accomplishments have resulted from these coordinated efforts and through external partnerships.

Major Accomplishments

Since the original HABHRCA legislation in 1998, several significant advances have greatly improved management. Many of these accomplishments are described in the five HABHRCA reports that were submitted to Congress. Rather than list every accomplishment, I will focus on recent outstanding achievements.

Harmful Algal Blooms

In the last few years, HAB prediction and forecasting has been extended to new areas and shown great promise in providing early warning to public health and resource managers. In most cases, the ability to provide HAB forecasts is the outcome of years of research efforts focused on the causes of HABs. Examples of regional HAB forecasting include:

- In Florida, the operational forecast system has issued over 500 forecasts since September 2004. These include the critical 2005 year, when *Karenia brevis* blooms (Florida red tide) struck three regions of Florida, on both the east and west coast, and produced anoxia on the Florida shelf for the first time in over 30 years. This forecast was made operational by strong NOAA-wide coordination, particularly between the National Ocean Service and National Environmental Satellite, Data, and Information Service, which processes and analyzes NASA MODIS satellite data through its Coast Watch Program.
- In Texas, an operational forecast began in September 2010 (following a several year demonstration). This system added new modeling capabilities which increase the adaptability of the forecasts. Previously, managers could respond only in a bay with a reported problem, and had no information as to which other bays were at greatest risk. The new forecasts provide this information to better target sampling and response. This is particularly important given the appearance in 2008 and 2010 of toxic blooms of *Dinophysis,* which were previously unknown in this area.
- In the Gulf of Maine, NOAA-funded researchers have issued seasonal advisories every year since 2008. Each spring they predict the severity and extent of blooms of *Alexandrium fundyense*, the New England HAB organism that produces a potent neurotoxin, which accumulates in shellfish and can cause human illness and death. That prediction provides state resource and public health managers time to prepare for the intensive monitoring required to protect public health and assists shellfish harvesters and processors in making business decisions. Weekly forecasts of *Alexandrium* distributions, based on models and weather forecasts, are also provided to state and local shellfish and public health managers around the Gulf of Maine via a listserv.
- In western Lake Erie, NOAA scientists have developed a Forecast System for cyanobacterial blooms starting in 2008. These blooms of the cyanobacterial HAB *Microcystis* have been recurring each summer for over 10 years, with particularly severe blooms in 2003, 2009, and 2010. The blooms are a significant expense for public water suppliers, and a potential human health risk through recreational use. In 2010, over 150 resource managers and local decision makers received the weekly demonstration forecasts of bloom location and intensity based on a sophisticated combination of satellite imagery from the European Space Agency (ENVISAT-1), circulation models, water analysis and meteorological data. In early October, NOAA determined that the bloom had ended, allowing Ohio to safely end sampling and analysis of water. (http://www.glerl.noaa.gov/ res/ Centers/HABS/lake_erie _hab/ lake_erie_hab.html)

- In Chesapeake Bay, a novel forecasting technique using a hybrid approach of water quality modeling and statistical techniques to predict HABs is nearing operational status. This forecasting tool also holds promise for other important applications such as pathogens and fish habitat which have been difficult to predict using other methodologies.
- Along the Washington coast, a toxic diatom, *Pseudo-nitzschia*, sometimes blooms and is transported to beaches where razor clams are harvested recreationally and by tribes. When exposed to such blooms, the clams accumulate the toxin, which can result in illness and death if the clams are eaten by humans. NOAA-funded scientists have improved early warning of *Pseudo-nitzschia* blooms by determining how winds move HABs from their source region to coastal beaches. Since 2008, these scientists have issued an interactive HAB Bulletin that managers from the Washington state Departments of Health and Fish and Wildlife use to determine, well in advance of openings, whether shellfish toxin levels will require closures. Managers can communicate this knowledge to harvesters and owners of coastal businesses catering to harvesters to minimize impacts.

Detection is a critical first step in protecting human health, as it is not possible to predict and respond to a problem that cannot be easily quantified or tracked. Many new methods of detecting HAB cells and toxins have been developed, tested, and in some cases put into routine use for a variety of purposes.

- State, local, and tribal shellfish and public health managers need quick tests that can be used for cheap and rapid screening for toxicity in many shellfish samples. NOAA-sponsored state, federal, and academic partnerships have demonstrated that new quick tests are reliable for screening large numbers of samples to rapidly assess the presence of HAB toxins in both shellfish and seawater and helped incorporate the new protocols into existing shellfish monitoring programs. States that routinely employ advanced HAB toxin screening tools include Washington, Oregon California, Florida, and Maine, and it is part of the screening method for the Shipboard/Dockside Screening Protocol for shellfish harvesting in Federal waters of Georges Bank.
- In Chesapeake Bay, new molecular techniques for detection of harmful algal species developed through the competitive HABHRCA programs are now in routine use by state agencies responsible for protecting resources and public health. These programs also allowed the state of Maryland to develop a unique and highly successful "Eyes on the Bay" website to display and communicate the latest information on HABs, hypoxia, and other observations in Chesapeake Bay. (http://mddnr. chesapeakebay. net/eyesonthebay/habs.cfm)
- For the last three years, a new instrument, developed and maintained with NOAA funding and located at Port Aransas Pass in Texas, has provided early warning of HAB outbreaks, resulting in closures of oyster harvesting before there were any human health impacts. One of the species detected has never before caused problems in the US, although it is common in Europe and there was no routine monitoring in place for that organism.

NOAA is addressing gaps in our understanding of the causes of HABs and responding to emerging HAB issues.

- Ciguatera fish poisoning (CFP) is the most common HAB-caused seafood illness in tropical and subtropical areas of the world. The incidence is increasing, perhaps linked to anthropogenic causes, such as overfishing, eutrophication and global warming. Economic impacts in the US due to human illness, which are believed to be hugely under-reported, are estimated to be $21M/yr.[9] The causative toxins were thought to come from one HAB species but NOAA scientists have determined that the difficulty in predicting CFP outbreaks is because there are multiple species of varying toxicity. Studies are underway to understand what controls the distribution and toxicity of these species in order to allow public health managers to minimize the incidence of this illness.
- Fish-killing algae in the Pacific Northwest have been shown to have severe economic impacts on mariculture. In addition, these algae might be a major factor in controlling the size of some wild salmon runs. NOAA scientists and NOAA-funded scientists are trying to identify the very ephemeral toxin and determine the causes of the blooms in order to develop protocols to minimize impacts on mariculture.

NOAA has always funded research on novel HAB mitigation and control measures through its existing HAB research programs. However, as our understanding of the causes of HABs has improved, more opportunities for preventing and controlling HABs have become available. In 2009, NOAA announced the establishment of a new Prevention, Control, and Mitigation of HABs Program, which is described above. The first projects were funded in 2010, and involve a diverse array of approaches on which we will be reporting at a later date.

Hypoxia

Through its HABHRCA-authorized hypoxia programs, NOAA has provided the research foundation upon which management of the "dead zone" in the Gulf of Mexico is based as described in the Mississippi River/Gulf of Mexico Watershed Nutrient Task Force Action Plan.[10] Ongoing targeted regional research is furthering our understanding of impacts on fisheries and local economies and filling gaps in our understanding of the factors driving the size and location of the hypoxic zone, including climate change. NOAA also forecasts and tracks the extent of hypoxia each year utilizing a number of internal and external programs and in concert with other federal agencies. This information is vital to support the Task Force's adaptive management approach to addressing this major coastal problem.

NOAA has collaborated closely with the US Environmental Protection Agency, the US Department of Agriculture, and other federal and state agencies in developing science-based management strategies to reduce nutrient pollution contributing to the Gulf of Mexico hypoxic zone. The Assistant Secretary of Commerce for Conservation and Management, Dr. Larry Robinson, sits on the interagency Mississippi River/Gulf of Mexico Watershed Nutrient Task Force, and NOAA also plays a leading role on the Task Force's Coordinating Committee.

Table 1. Major HAB organisms causing problems in US marine systems, their major toxins (if characterized), their direct acute impacts to humans and ecosystem health, and regions of the US that have been impacted by these HAB organisms. 'Not characterized' indicates that toxins have been implicated but not characterized

Organisms	Toxins	Acute Human Illness*	Direct Ecosystem Impacts	Impacted Regions
Alexandrium spp.	Saxitoxins	Paralytic Shellfish Poisoning	Marine mammal mortalities	Northeast, Pacific Coast, Alaska
Aureococcus anophagefferens (Long Island Brown Tide)	Not characterized	--	Shellfish mortality, seagrass die-off	Northeast, Mid-AtlanticCoast
Aureoumbra lagunensis (Texas Brown Tide)	Not characterized	--	Seagrass die-off	Gulf of Mexico (Texas)
Dinophysis	Okadaic Acid	Diarrhetic Shellfish Poisoning	--	Gulf of Mexico, possibly New England and Pacific Coast
Gambierdiscus spp., *Prorocentrum* spp., *Ostreopsis* spp.	Ciguatoxin, Gambiertoxin, and Maitotoxin	Ciguatera Fish Poisoning	--	Gulf of Mexico (Florida, Texas), Hawaii, Pacific Islands, Puerto Rico and US Virgin Islands
High biomass bloom formers	†	--	Low dissolved oxygen, Food web disruption	All regions
Karenia spp.	Brevetoxins	Neurotoxic Shellfish Poisoning, Acute respiratory illness	Fish kills, mortalities of other marine animals	Gulf of Mexico, South-Atlantic Coast
Karlodinium spp.	Karlotoxins	--	Fish kills	Mid- and South-Atlantic Coast, Gulf of Mexico (Alabama, Florida)
Macroalgae	‡	--	Low dissolved oxygen, seagrass and coral overgrowth and die-off, beach fouling	All regions
Marine Cyanobacteria (CyanoHABs) (*Lyngbya* spp)	Lyngbyatoxins	Dermatitis	Seagrass and coral overgrowth and die-off, beach fouling	Gulf of Mexico and South-Atlantic Coast (FL), Hawaii and Pacific Territories
Pfiesteria spp.	Free radical toxin, others not characterized	--	Fish kills	Mid- and South-Atlantic Coast
Pseudo-nitzschia spp.	Domoic Acid	Amnesic Shellfish Poisoning	Mortality of seabirds and marine mammals	Pacific Coast, Alaska, Gulf of Mexico, Northeast, Mid-AtlanticCoast
Pyrodinium bahamense	Saxitoxins	Puffer Fish Poisoning	--	South-Atlantic Coast (Florida)
Some raphidophytes (e.g., *Heterosigma akashiwo, Chattonella spp.*)	Brevetoxins (*Chattonella*), other icthyotoxins not characterized	--	Fish kills	Pacific Coast (Washington), Mid-Atlantic Coast

*This table only captures the major acute human illnesses associated with these HAB species. Other, less severe acute health effects, such as skin irritation, may occur with some of these HAB groups. Chronic effects, such as tumor promotion, can also occur. A table of short- and longterm health effects is given in12.

†Some high biomass bloom formers may produce toxins.

‡Some macroalgae have been shown to produce bioactive compounds, such as dopamine and dimethylsulfoniopropionate (DMSP), which may have direct ecosystem effects (Van Alstyne et al. 2001).

The Task Force released the 2008 Gulf Hypoxia Action Plan, which reaffirmed the goal of reducing the hypoxic zone and suggested 45 percent reductions of both total nitrogen and phosphorus.

NOAA-funded research through the NGOMEX program has demonstrated that widespread reproductive impairment (reduced ovarian and testicular growth in adults, and decrease in hatching success and larval survival) occurs in a common marine fish, Atlantic croaker, in the hypoxic zone west of the Mississippi River. More recently, the actual molecular mechanism behind the reproductive impairments in fish was identified, adding to a growing body of evidence that non-lethal hypoxia impacts pose long-term threats to living resource populations in hypoxic zones.

Other studies are determining the impacts of hypoxic zones on the economics of shrimp fisheries, and on populations of other ecologically and commercially valuable fisheries.

NOAA-funded researchers are providing predictive modeling tools to resource and water quality managers in Narragansett Bay in Rhode Island to help mitigate hypoxia events, which have led to major fish kills and resulted in State nutrient reduction criteria[11] for waste water treatment plants (WWTPs). These predictive modeling tools will provide alternative management options for WWTPs (such as relocation of outfall pipes to locations where outward currents would speed nutrients out of the ecosystem) and will generate ecological impact scenarios for various nutrient loading estimates, thereby helping to determine allowable nutrient loadings for WWTPs into local rivers that drain into Narragansett Bay.

CONCLUSION

Thank you for this opportunity to update you on NOAA's HAB and hypoxia programs. Over the last twelve years, we have made enormous progress in understanding the causes and consequences of HABs and hypoxia. This has led to direct and significant improvements in HAB and hypoxia management which have, in turn, protected public health and vital economic interests.

The Administration supported reauthorization of HABHRCA in the last Congress and continues to support reauthorization in the 112th Congress. We just recently received the draft legislation. We will review it, along with other interested Departments and agencies.

We would appreciate an opportunity to comment before the Subcommittee considers the legislation.

End Notes

[1] Lopez, C.B., Dortch, Q., Jewett, E.B., Garrison, D. 2008. Scientific Assessment of Marine Harmful Algal Blooms. Interagency Working Group on Harmful Algal Blooms, Hypoxia, and Human Health of the Joint Subcommittee on Ocean Science and Technology. Washington, D.C., 62 pp.

[2] Lopez, C.B., Jewett, E.B., Dortch, Q., Walton, B.T. Hudnell, H.K. 2008. Scientific Assessment of Freshwater Harmful Algal Blooms. Interagency Working Group on Harmful Algal Blooms, Hypoxia, and Human Health of the Joint Subcommittee on Ocean Science and Technology. Washington, D.C., 65 pp.

[3] Committee on Environment and Natural Resources. 2010. Scientific Assessment of Hypoxia in US Coastal Waters. Interagency Working Group on Harmful Algal Blooms, Hypoxia, and Human Health of the Joint Subcommittee on Ocean Science and Technology. Washington, DC.., 164 pp.

[4] Hoagland, P., and Scatasta, S. 2006. The Economic Effects of Harmful Algal Blooms. In E Graneli and J Turner, eds., Ecology of Harmful Algae. Ecology Studies Series. Dordrecht, The Netherlands: Springer-Verlag, Chap. 29.

[5] Jin, D., Thunberg, E., and Hoagland, P. 2008. Economic Impact of the 2005 Red Tide Event on Commercial Shellfish Fisheries in New England. Ocean and Coastal Management. 51(5): 420-429.

[6] GEOHAB, 2006. Global Ecology and Oceanography of Harmful Algal Blooms, Harmful Algal Blooms in Eutrophic Systems. P. Glibert (ed.). IOC and SCOR, Paris and Baltimore, 74, pp.

[7] Heisler, J., Glibert, P.M., Burkholder, J.M., Anderson, D.M., Cochlan, W., Dennison, W.c., Dortch, Q., Gobler, C.J., Heil, C.A., Humphries, E., Lewitus, A., Magnien, R., Marshall, H.G., Sellner, K., Stockwell, D.A., Stoecker, D.K., and Suddleson, M. 2008. Eutrophication and Harmful Algal Blooms: A Scientific Consensus. Harmful Algae 8(1): 3-13.

[8] Jewett, E.B., Lopez, C.B., Dortch, Q., Etheridge, S.M., Backer, L.C. 2008. Harmful Algal Bloom Management and Response: Assessment and Plan. Interagency Working Group on Harmful Algal Blooms, Hypoxia, and Human Health of the Joint Subcommitttee on ocean Science and Technology. Washington, DC., 76 pp.

[9] Anderson, D. M., Hoagland, P., Kaoru, Y. & White, A. W. 2000. Estimated Annual Economic Impacts from Harmful Algal Blooms (HABs) in the United States. Technical Report WHOI 2000-11. Woods Hole Oceanographic Institution, 99 pp.

[10] Mississippi River/Gulf of Mexico Watershed Nutrient Task Force. 2008. Gulf Hypoxia Action Plan 2008 for Reducing, Mitigating, and Controlling Hypoxia in the Northern Gulf of Mexico and Improving Water Quality in the Mississippi River Basin. Washington, DC.

[11] RIDEM. 2005. Plan for managing nutrient loadings to Rhode Island waters (http://www.dem.ri.gov/pubs/nutrient.pdf). 17pp. and RIDEM. 2006. Water quality regulations. Office of Water Resources, Department of Environmental Management, State of Rhode Island and Providence Plantations. [http://www.dem.ri.gov/programs/benviron/ water/ quality /surfwq /index.htm]

[12] Ramsell, J.S., Anderson, D.M., and P.M. Glibert (eds.). 2005. Harmful Algal Research and Response: A National Environmental Science Strategy 2005-2015. Ecological Society of America, Washington, D.C., 96 pp.

In: Assessing the Hypoxia Threat in U.S. Coastal Waters ISBN 978-1-62081-303-4
Editors: William Jacobs and Steve Frederickson

Chapter 3

THIS IS TESTIMONY OF RICHARD GREENE, CHIEF, ECOSYSTEM DYNAMICS AND EFFECTS BRANCH, GULF ECOLOGY DIVISION, US ENVIRONMENTAL PROTECTION AGENCY. HEARING ON "THE HARMFUL ALGAL BLOOMS AND HYPOXIA RESEARCH AND CONTROL AMENDMENTS ACT OF 2011"*

Good afternoon Chairman Harris, Ranking Member Miller, and other members of the Subcommittee. My name is Richard Greene, with EPA's Office of Research and Development (ORD).

For the last 13 years, I've served as Chief of the Ecosystem Dynamics and Effects Branch at the Gulf Ecology Division, within the National Health and Environmental Effects Research Laboratory.

I have a Ph.D. in Oceanography and over the last 10 years have been the ORD lead for Gulf of Mexico hypoxia research and estuarine nutrient research in the Gulf. Although I serve in an ecology research Division along the Gulf of Mexico, this testimony addresses programs across EPA offices and laboratories relevant to the focus of this hearing. It is a pleasure to be here with you today to discuss the EPA's research relating to harmful algal blooms and hypoxia.

HARMFUL ALGAL BLOOMS AND HYPOXIA – THREATS TO HUMAN HEALTH AND ECOSYSTEMS

Toxic or otherwise harmful algal blooms and hypoxia, or low dissolved oxygen, represent significant and continuing threats to freshwater, estuarine and coastal ecosystems, aquatic life and human health. Scientific understanding of the causes and impacts of harmful algal blooms

* This is an edited, reformatted and augmented version of testimony given on June 1, 2011 before the Subcommittee on Energy and Environment, Committee on Science, Space, and Technology, US House of Representatives.

and hypoxia on aquatic ecosystems and human health has progressed over the last 5-10 years. However, there is still much to be learned to improve our ability to predict when and where those events will occur, the specific impacts they will have on human health and aquatic ecosystems, and how best to prevent, control or mitigate those problems. We know that by-and-large many HAB and hypoxia events are inextricably linked to nitrogen and phosphorus pollution. However, we need to improve the science supporting the development of sustainable solutions for controlling and reducing nutrient pollution, HABs and hypoxia, and protecting water resources and human health. EPA, working with NOAA, and other federal, state, and private sector partners, is committed to that goal.

Nationally, nitrogen and phosphorus pollution is one of the top causes of water quality impairments. EPA's National Aquatic Resource Surveys (NARS) show that of the stressors assessed, nitrogen and phosphorus are the most pervasive in the nation's wadeable streams, with more than 200,000 stream miles showing high concentrations (those greater than 95 percent of the regionally-relevant least-disturbed reference condition)[1].

The NARS also report that an estimated four million lake acres showed high concentrations of phosphorus, and 1.9 million acres showed high concentrations of nitrogen[2]. Streams and lakes with high levels of nitrogen and phosphorus were about two times more likely to have poor biological health[3] . For streams, biological health was determined by evaluating the health of macroinvertebrate communities compared with least-disturbed, regionally-relevant, reference conditions.

In lakes, biological condition was determined by analyzing the condition of zooplankton and phytoplankton communities using an observed/expected model. The ecological impact of excess nutrients on our waters includes harmful algal blooms.

The recent NARS lakes assessment found microcystin (an algal toxin that can harm humans, pets, and wildlife) present in about one-third of lakes and at levels of concern in one percent of lakes based on World Health Organization recreational exposure guidelines[4]. Although there are relatively few documented cases of severe human health effects , exposure to cyanobacteria or their toxins may produce allergic reactions such as skin rashes, eye irritations, respiratory symptoms, and in some cases gastroenteritis, liver and kidney failure, or death.

The most likely exposure route for humans is through accidental ingestion or inhalation during recreational activities, though cyanotoxins are also potentially a cause for concern in drinking water.[5]

Under Section 303(d) of the Clean Water Act, states develop lists of impaired waters every two years and are then required to develop clean-up plans, also known as Total Maximum Daily Loads (TMDLs), for those waters. A TMDL is a calculation of the maximum amount of a pollutant that a waterbody can receive and still safely meet water quality standards. Currently, more than 15,800 waters have nitrogen- or phosphorus-related impairments. States or EPA has developed more than 8,000 nitrogen- or phosphorus-related TMDLs for more than 5,000 of these waters.

Grand Lake St. Marys stands out as one recent example of the potentially significant and far-reaching costs associated with the human health, economic, recreational and ecosystem impacts of nitrogen- and phosphorus-contaminated waters. Grand Lake St. Marys, Ohio's largest inland water body, has suffered from increasing nitrogen and phosphorus loading from farm runoff, failing septic systems, and fertilizer applied to lawns. As a result, the lake has experienced massive blooms of toxic cyanobacteria, which have led to the death of fish, birds, and dogs, and illnesses of at least seven people. The State of Ohio has issued fish consumption, recreational use, and health warnings, including "no contact" and "algal bloom" advisories.[6]

In addition to Grand Lake St. Marys, freshwater HABs or the cyanoHABs have now been documented in at least 35 states and at least 18 states now have some type of CyanoHAB research or response program.

EPA continues to evaluate the human health implications of cyanoHABs and the toxins they produce in drinking water. Cyanotoxins have been included in all three Candidate Contaminant Lists (CCL) published so far pursuant to the Safe Drinking Water Act. EPA must periodically publish this list of contaminants and decide whether to regulate at least five or more contaminants on the list.

A drinking water CCL is the primary source of priority contaminants from which EPA makes decisions about whether regulations are needed. The contaminants on the list are known or suspected to occur in public water systems but are currently unregulated by existing drinking water standards. The Agency included cyanotoxins as a group in the most recent CCL (CCL 3, October 8, 2009) and focuses research and data collection on three algal toxins: Anatoxin-a; Microcystin-LR; and Cylindrospermopsin.

Epa Research

Freshwater HABs

EPA's National Aquatic Resource Surveys are beginning to contribute significant information necessary to evaluate the extent and impact of harmful algae, nutrients, and other key indicators on ecological condition and potential human health risks. The 2009 report on the National Lakes Assessment included three indicators with respect to the condition and safety of recreational water use: 1) microcystin - a common algal toxin, 2) cyanobacteria - the group of unicellular or filamentous algae, some of which produce algal toxins, and 3) chlorophyll-a - a measure of all algae present.

The study reported that microcystin was present in about 30% of lakes and at levels of concern in about 1% of lakes, based on World Health Organization thresholds of risk. While the survey results are a good start in our understanding, much more is to be learned about algal toxins in lakes. For example, it is currently unknown how well microcystin occurrence correlates with the occurrence of other classes of cyanotoxins that were not measured, or the associated human health risks.

In addition, there are relatively few documented cases of severe human health effects associated with exposure to cyanobacteria or their toxins.

In addition, ORD is conducting a pilot study to assess human exposure and effects, and potential developmental toxicity associated with the cyanotoxins microcystin, cylindrospermopsin and anatoxin-a in drinking water systems in the southeastern United States.

The pilot study is to determine whether a new blood serum assay for microcystin exposure can reliably detect low levels of human exposure to the toxin. A second study of developmental toxicity uses human placenta cells in culture to determine whether microcystins disrupt normal placental formation for pregnancy maintenance.

Gulf of Mexico Hypoxia

ORD coordinates and collaborates research in the northern Gulf hypoxia zone with NOAA and other federal, state and academic organizations. ORD has ongoing research and modeling efforts in the northern Gulf of Mexico to assess and predict: the relationships between nutrient loads and hypoxia; the physical, chemical, and biological processes regulating dissolved oxygen dynamics in the Gulf; and the effects of nutrient load reduction scenarios on hypoxia. ORD has partnered with Naval Research Laboratory modelers to develop integrated water quality simulation modeling tools that will improve our ability to evaluate the effectiveness of nutrient load reductions on Gulf hypoxia. The research and modeling efforts underway fill important research gaps identified in 2007 by EPA's Science Advisory Board state-of-the science evaluation regarding Gulf hypoxia and are critical to the information needs and goals of the Mississippi River/Gulf of Mexico Watershed Nutrients Task Force (Gulf Hypoxia Task Force). In 2009, ORD published research on multiple regression models that described the size of the Gulf hypoxic zone based on river discharge and nitrogen and phosphorus concentrations. Those results supported the need for a dual nutrient management strategy - reductions in both nitrogen and phosphorus loads - to achieve the goal of reducing Gulf hypoxia. Equally significant were results of model predictions demonstrating that substantial and sustained nitrogen and phosphorus reductions will be needed before it will possible to discern statistically significant reductions in hypoxic area against the background of natural variability. ORD scientists, as well as NOAA and NOAA-funded academic groups are working on parallel efforts to develop multiple 3D numerical simulation models for the northern Gulf system that link nutrient inputs, physical circulation processes, and ecological and water quality responses. EPA, NOAA and the scientific community consider the development of multiple models and modeling approaches as offering many advantages compared to a single model or modeling approach for addressing complex problems like the Gulf of Mexico hypoxia. All the research groups met recently in Mississippi at a NOAA-sponsored workshop to report on progress of the modeling efforts. The groups are about a year or two away from being able to run 3D model simulations and examine the effects of nutrient load reduction scenarios on dissolved oxygen dynamics and ultimately, the size, frequency, and duration of hypoxia in the northern Gulf.

EPA PATICIPATION IN INTERAGENCY TASK FORCES AND WORKGROUPS

Harmful Algal Bloom and Hypoxia Task Force

ORD is an active participant in the Interagency Working Group on HABs, Hypoxia, and Human Health (IWG-4H) within the National Ocean Council, which is led by NOAA and which, among other responsibilities, implements the reporting requirements of the Harmful Algal Bloom and Hypoxia Research and Control Act (HABHRCA) of 2004. In these efforts, ORD staff were co-authors on the *Scientific Assessment of Hypoxia in US Coastal Waters,* released in 2010, and the *Scientific Assessment of Freshwater Harmful Algal Blooms*, released in 2008, through IWG-4H.

EPA has had a long-standing collaboration with NOAA through the Interagency Ecology and Oceanography of Harmful Algal Blooms Program, authorized by HABHRCA in 1998 and 2004. A Memorandum of Understanding allowed the participating agencies, EPA, NOAA, NSF, NASA, and ONR, to fund competitive research on the causes and impacts of HABs and to develop methods of detection, prevention and control. EPA funded nearly 30 projects between 1997 and 2006 several of them joint efforts with NOAA.

Harmful algal blooms are of concern in the Great Lakes and other waters because of their toxicity and impact on human and ecosystem health. A particularly toxic species is present in Western Lake Erie and Saginaw Bay (Lake Huron) -- two areas of the Great Lakes that typically have significant cyanobacterial blooms. These blooms cause fouling of the beaches and shoreline, economic and aesthetic losses, taste and odor problems in drinking water, and direct risks to human, fish and animal health. EPA's Great Lakes National Program Office funds research on harmful algal blooms research and coordinates with NOAA's Center of Excellence for Great Lakes and Human Health (CEGLHH). Grants associated with nutrient related controls, management, and restoration have been a significant area of emphasis in the Great Lakes Restoration Initiative. Restoration related activities are under way as part of grant-funded projects at several sites across the Great Lakes.

Mississippi River/Gulf of Mexico Watershed Nutrients Task Force

In addition to EPA's participation in the Federal interagency Task Force on HABs and Hypoxia, EPA OW co-chairs the Mississippi River/Gulf of Mexico Watershed Nutrient Task Force (Gulf Hypoxia Task Force) which is comprised of 17 state and federal agencies. The Gulf Hypoxia Task Force provides a forum for state water quality and agriculture agencies to partner on local, state, and regional efforts to mitigate nutrient loading, encouraging a holistic approach that takes into account upstream sources and downstream impacts. The Task Force's goal of reducing the size of the Gulf hypoxic zone to a five-year running average of 5,000 km^2 is a very challenging commitment.

In its most recent *Gulf Hypoxia Action Plan* (2008), the Gulf Hypoxia Task Force emphasized its commitment to work with states to develop nutrient reduction strategies and increase accountability, and both remain top priorities of the current Gulf Hypoxia Task Force leadership. To facilitate nutrient strategy development, the Gulf Hypoxia Task Force produced a State Nutrient Reduction Strategy Report in September 2010 that identifies essential strategy components and potential federal funding sources. As recommended in that report, Gulf Hypoxia Task Force federal and state members, including EPA, have assisted in organizing and providing technical and funding support for two nutrient reduction strategy workshops, the second of which is occurring in mid-June in Columbus, Ohio. A Gulf Hypoxia Task Force working group is developing a proposed set of "indicators of progress" to measure progress towards addressing nitrogen and phosphorus pollution in the Mississippi and Atchafalaya River Basin (MARB) and reducing the size of the Gulf hypoxic zone.

More broadly, EPA has asked interested and willing states to join the Agency, other federal partners, and stakeholders and work collaboratively to achieve substantial near-term reductions of nitrogen and phosphorus pollution, using a transparent and accountable action framework, while some states continue to develop numeric criteria for nitrogen and phosphorus pollution to provide a clearly measureable and objective basis for longer-term reduction strategies.[7]

ORD provides technical support to OW for Gulf Hypoxia Task Force activities and also participates in the Task Force Coordinating Committee. ORD collaborates and coordinates with NOAA and other organizations thru the Gulf of Mexico Hypoxia Research Coordination Workshop series (sponsored by NOAA) which seeks to coordinate monitoring and modeling activities in the Gulf hypoxic zone.

CONCLUSION

In conclusion, EPA has made progress in understanding and addressing harmful algal blooms, hypoxia, and the broader issues of nutrient pollution in the US, and there is much more to do. EPA programs are targeting the causes and their impacts, working with states and federal partners to identify and protect healthy watersheds and their receiving waters and restore impaired waters. These efforts will improve management of nutrients, HABs and hypoxia, and help create safe and sustainable water resources for the future generations. We look forward to working with the Committee in the future.

Thank you for the opportunity to be here today, and I will be happy to answer your questions.

End Notes

[1] Wadeable Streams Assessment: A Collaborative Survey of the Nation's Streams EPA 841-B-06-002 December 2006. Chapter 2 – see page 35 -http://water.epa.gov/type/rsl/monitoring / streamsurvey/upload/2007_5_9_ streamsurvey 05_chap2a 5-2-07.pdf

[2] Note – The NLA reported information as the number and/or percent of lakes. See - US Environmental Protection Agency (USEPA). 2009. National Lakes Assessment: A Collaborative Survey of the Nation's Lakes. EPA 841-R-09-001. US Environmental Protection Agency, Office of Water and Office of Research and Development, Washington, D.C. See Chapter 3 - http://water.epa.gov/type/lakes/upload/nla_newchapter3.pdf To make the lake statement more consistent with the streams statement (e.g., using miles and acres rather than miles and percent of lakes), ORDCorvallis calculated the area associated with the % of lakes presented in the NLA document on page 26. The area figures were emailed by Steve Paulsen, EPA-ORD to Sarah Lehmann, EPA-OW on 2-16-2011.

[3] Wadeable Streams Assessment: A Collaborative Survey of the Nation's Streams EPA 841-B-06-002 December 2006. See Chapter 2 – page 49. http://water.epa.gov/type/rsl/monitoring/ streamsurvey/upload/2007_5_14_ streamsurvey 06_chap2b_5-2-07.pdf

[4] US Environmental Protection Agency (USEPA). 2009. National Lakes Assessment: A Collaborative Survey of the Nation's Lakes. EPA 841-R-09-001. US Environmental Protection Agency, Office of Water and Office of Research and Development, Washington, D.C. See Chapter 4 - http://water.epa.gov/type/lakes/upload/ nla_newchapter4.pdf

[5] IBID

[6] Spencer Hunt, New Tests find Grand Lake St. Marys Safe, The Columbus Dispatch, 30 Oct. 2010. One Algal Toxin Advisory Removed; Others Continue, Ohio EPA. 29 Oct. 2010. Ohio EPA http://www.epa.state. oh.us/pic/glsm_algae.aspx

[7] N. Stoner, March 16, 2011. Working in Partnership with States to Address Phosphorus and Nitrogen Pollution through Use of a Framework for State Nutreint Reductions. US Environmental Protection Agency, Washington D.C. See http://water.epa.gov/scitech/ swguidance/standards/criteria/nutrients/upload/ memo_ nitrogen_framework.pdf

INDEX

C

D

E

F

G

M

N

O

P

Q

R

S

T

U

V

W

Y

Z